The Q...
for Absolute ...

The Quest for Absolute Zero

the meaning of low temperature physics

K. Mendelssohn

Second Edition with S.I. Units

TAYLOR & FRANCIS LTD

London 1977

First edition published in 1966 by Weidenfeld & Nicolson,
11 St. John's Hill, London.

Second edition published in 1977 by Taylor & Francis Ltd,
10–14 Macklin Street, London WC2B 5NF.

ISBN 0 85066 119 6

Printed and bound in Great Britain by
Taylor & Francis (Printers) Ltd,
Rankine Road, Basingstoke, Hampshire.

This second edition is dedicated
to the memory of Heinz London

Contents

Introduction

Ten years have passed since this book was first published and in the meantime it has appeared in thirteen different languages, indicating an appreciation that is most gratifying to the author. The British and American paperback editions have been sold out and there has been a mounting demand on the part of university teachers to make the book again available. However, I feel that a straight reprinting would hardly have been justified since it cannot do justice to the subject which in scope and direction has changed vastly in the past decade. The historical and general aspects have naturally not required alterations and the first seven chapters have remained essentially unaltered. Matters are, however, different when we have to deal with sections of the book which ten years ago represented the latest state of the art.

In chapter 8, dealing with magnetic cooling, note has had to be taken of the development of the nuclear cooling method and particularly of its extension into the field of nuclear refrigeration. The next chapter, on superconductivity has had to be expanded considerably to take account of some confirmatory experiments on flux quantisation and its part in the elucidation of type II superconductivity. In addition, there has been much support for the BCS theory which can now be regarded as the firmly established interpretation of superconductivity. Some indication was given in the first edition of the use of materials which remain superconductive to very high magnetic fields. The subsequent development in the manufacture and use of these materials has taken place at truly breathtaking speed and within the last decade superconductive technology has progressed at an ever increasing rate. Coupled with cryogenic requirements of space technology, the world of low temperature physics has been completely transformed from the narrowly circumscribed scope of academic investigation to a world of large-scale enterprise. It is obviously quite unrealistic to close one's eyes to the completely new direction that low temperature work is taking. Trying to encompass these technological developments by simply enlarging the chapter 9, originally devoted to *all* aspects of superconductivity, was quite impossible, and so an entirely new chapter on the various facets of superconductive technology has been added. Extensive additions were also required in the chapter, now number 11, on superfluidity. The recent discovery of superfluidity of liquid ^{3}He in the millikelvin range has provided the one outstanding new feature in low temperature physics in the last decade. It is also of fundamental significance for the basic understanding of the phenomena of superconductivity and superfluidity as a universal feature of the behaviour of matter at the lowest temperatures. In addition, note had to be taken of new cooling methods in this range.

While these last interesting discoveries testify to the continued appeal of fundamental research in low temperatures, the recent technological significance of the field has brought about a distinct change in emphasis from the pure to the applied aspects. In fact, compared with the work in the late thirties and its continuation during the post-war years, the volume of basic research has not increased. There has been a distinct lessening of interest in pure research, mainly because, apart from the exciting development in the ^{3}He field, hardly any new fundamental problems have arisen. On the other hand, many capable and imaginative young scientists have been strongly attracted to the creation of a new and exciting technology within a few degrees above absolute zero, undreamt of in the heyday of fundamental research twenty-five years ago.

It is perhaps significant that the important centres of basic research in those days have not become the home of the new applied work. A typical example is the Clarendon Laboratory at Oxford, once one of the foremost institutions in the field of superconductivity and superfluidity where now neither of these subjects is studied. Similar changes in interest have occurred in others of the old established laboratories and the new fundamental work on ^{3}He has shifted to universities such as Cornell and San Diego in the U.S.A., Otaniemi in Finland and Manchester and Sussex in England. The leaders of these schools were, as often as not, trained in the old institutions who took their interest with them to new ground. A good deal of this movement can be ascribed to personal reasons. The pioneers of the thirties have now retired and those taking their places have understandably lacked the adventurous frontier spirit that characterised the heyday of low temperature research. In what appears to be a contracting subject, their laboratories have taken refuge in conscientious, if less exciting, routine work of general solid state physics. The pioneers of the present generation have moved into the neighbouring field of low temperature technology which offers ample scope for new ideas and achievements on an unprecedented scale. It lies in the nature of this applied work that it must be conducted not in a cloistered academic atmosphere but in close co-operation with engineers and technicians who are acquainted with production processes. Their work has therefore moved to technical universities and above all to laboratories directly connected with government and industry.

Thus the decade separating the present edition from the first publication of this book is characterised not only by new basic discoveries but primarily by a fundamental shift in the direction of low temperature work. At present cryogenic research is undergoing a change of emphasis and it can be predicted with confidence that in a few years' time it will have gravitated from the university laboratory to centres of technological development. In this we are only at the very beginning and what the future will bring, nobody can say at present. As far as fundamental low temperature research is concerned, we seem to be standing much closer to a completed subject than we did when the first edition of book was published.

The heyday of low temperature physics.
L. D. Landau in discussion with the author in Moscow, 1957.

1 Paris 1877

Our story opens in 1877. The date is Christmas eve and the scene is set at the Academy of Sciences in Paris. In the preceding week rumours of the impending announcement of an important discovery had thickened and the assembly was waiting impatiently while official business was being despatched in the ponderous manner of Learned Academies. Then, as a curtain raiser, the secretary, M. Dumas, read a quotation from the works of Lavoisier:

If the earth were taken into a hotter region of the solar system, say one in which the ambient temperature were higher than that of boiling water, all our liquids and even some metals would be transformed into the gaseous state and become part of the atmosphere. If, on the other hand, the earth were taken into very cold regions, for instance, to those of Jupiter or Saturn, the water of our rivers and oceans would be changed into solid mountains. The air, or at least some of its constituents, would cease to remain an invisible gas and would turn into the liquid state. A transformation of this kind would thus produce new liquids of which we as yet have no idea.

Almost a century had passed since the great chemist had written these prophetic words and all efforts to create in the laboratory Lavoisier's 'new liquids' had failed. The communication which was now to be read, and which announced the liquefaction of oxygen, was therefore a milestone on a new road, the road to the absolute zero of temperature.

The author of the communication was a newly elected corresponding member of the Academy, a mining engineer from Chatillon sur Seine by the name of Cailletet. Like many of his predecessors, he began to experiment with gases in the hope of obtaining liquefaction by subjecting them to high pressure. Acetylene (C_2H_2) was chosen for the first attempt because it had been suggested that at room temperature a pressure of about 60 atmospheres might be sufficient to turn it into a liquid. However, before this pressure was reached, the apparatus sprang a leak and the compressed gas escaped. Cailletet had been watching the strong-walled glass tube which contained the compressed acetylene when the mishap occurred. He noticed in the moment following the accident that a faint mist had formed in the tube only to disappear again immediately. The key to his subsequent achievement was Cailletet's immediate conclusion that the release of pressure had resulted in a sudden cooling of the acetylene and that he had witnessed the temporary condensation of a gas. His first suspicion was that the acetylene had been impure and that he had seen droplets of water. Berthelot's famous laboratory supplied him with very pure acetylene but when the experiment was repeated the mist appeared again. There could now be no doubt that acetylene had been liquefied and that Cailletet's mishap had provided him with a new technique for gas liquefaction.

Armed with this knowledge, he lost no time in attacking the most important problem in the field: the liquefaction of the atmospheric gases. He began with oxygen because it could be prepared fairly easily in a pure state. He compressed this gas to a pressure of about 300 atmospheres and cooled the strong-walled glass tube of his apparatus to -29°C by surrounding it with evaporated sulphur dioxide. When the pressure was suddenly released, he again observed a mist of condensing droplets and by a number of tests satisfied himself that it was not due to impurity. Oxygen had been liquefied and it was an account of this experiment which he submitted to the Academy.

However, another surprise was to come. Cailletet's paper had hardly been read when the Secretary announced that two days earlier, on 22 December, the Academy had received a telegram from a physicist in Geneva. It read:

> Oxygen liquefied to-day under 320 atmospheres and 140 degrees of cold by combined use of sulfurous and carbonic acid. RAOUL PICTET

This short message was augmented by an account, evidently sent in earlier and in anticipation of the result, of the method employed by Pictet. He had not found Cailletet's shortcut to success but had approached the liquefaction of oxygen by a series of steps, each representing the liquefaction of a different gas which has become known as a 'cascade', and which will be discussed later. The important point is that Pictet had reached the same goal as Cailletet, at essentially the same time but by a quite different method.

It was not an accidental coincidence of discoveries. The history of science provides us with many instances in which the same discovery was made independently by two or more people at roughly the same time. When we say 'independently', it really means that they did not crib from each other or look over each other's shoulder. These discoveries were never independent when the state of scientific knowledge as a whole is considered. Science does not proceed by a series of lucky accidents. It is an organic growth which inevitably will produce the discoveries required by each developmental stage. Cailletet would have passed over the observation of the mist if he had not been looking out for liquefaction. If he had failed to grasp its significance, somebody else would have noticed it soon enough. Even so, there was Pictet, arriving at the same result by another road. As we shall see later, although our story begins with a double fanfare on Christmas eve 1877, the stage was set for it and the grand opening had to come just about then.

The organic progress of science with the inevitability of its achievements at the right time has presented its practitioners with a thorny problem; that of priority. While everybody knows that when a discovery was made, it had to be made, a good deal of kudos is nevertheless attached to having made it first. One knows of a sufficient number of cases which come close enough to a neck-and-neck race, requiring a photo-finish. There are certain rules to the game which vary slightly from time to time and from country to country but which all centre around some official publication of the discovery. In France, at the time of Cailletet and Pictet, it was communication to the Academy which counted. By his telegram Pictet had secured the 22nd December, the day on which his experiment had succeeded, but what about Cailletet, whose paper was read only on the 24th? His decisive observation had been made on the 2nd December, and it was for purely personal reasons that no announcement had been made at the meetings of the 3rd, the 10th, or even the 17th. Cailletet had been in a quandary. He was up for election as a Corresponding Member of the Academy at its meeting on the 17th December and he wisely thought that it would be bad policy to announce a sensational result on or just before that date. The election was contested and it might look like an attempt at influencing the voters. On the other hand, there was no harm in letting the cat out of the bag provided it was not done in the Academy, and Cailletet had arranged for a demonstration of oxygen liquefaction before an invited audience of colleagues at the École Normale in Paris on Sunday, the 16th December. On the following day the Academy elected him with thirty-three votes to nineteen.

Now the important announcement could be made on the 24th, but there was Pictet's telegram of the 22nd. It seemed as if Cailletet had lost his priority. However, the story has a happy ending. On the 2nd December, the day of his crucial experiment in Chattilon-sur-Seine, Cailletet had written a letter giving a full account of the discovery to his friend, Sainte-Claire Deville in Paris which arrived there on the 3rd. Saint-Claire Deville had taken the letter immediately to the Permanent Secretary of the Academy, who had signed and sealed it with the date. Cailletet had now officially become the first to liquefy oxygen.

Pictet had said nothing in his first communication about the ultimate temperature of his experiment. Cailletet had estimated the temperature drop in his expansion as 200°C, which as a very rough guess was not too far wrong. At that stage it was not so much decrease in temperature which seemed significant as the fact that Lavoisier's prediction had been proved right in the laboratory. The air of the earth's atmosphere had been turned into a liquid, for, only a week after the memorable meeting of the 24th, Cailletet announced the liquefaction of nitrogen. He had been following up his success during the Christmas week.

Guillaume Amontons, who lived in the second half of the seventeenth century, seems to have been first in formulating the concept of an absolute zero of temperature. He was a late contemporary of Boyle and Mariotte who, again independently, had shown that the pressure of air increased in the same proportion as its volume is reduced when it is compressed. Amontons had lost his hearing as a child and devoted his life to the study of temperature and its measurement. In his endeavour to construct a reliable thermometer he used a volume of air, confined by a column of mercury as an indicator. He thus extended the work of Boyle and Mariotte by measuring the change in pressure of a given volume of air when its temperature is varied. He started his measurement at the boiling point of water and noted that equal drops in temperature resulted in equal decreases in the pressure of the air. From this he concluded that eventually on further cooling the air pressure should become zero at a finite degree of temperature which he estimated as −240°C. Since the pressure of the gas cannot become negative, it follows that there must exist a lowest temperature beyond which air, or any other substance, cannot be cooled. Amontons thus anticipated the work of Charles and also of Gay-Lussac in France, who a century later, and once more independently, formulated this law in a more rigorous form. They showed that the drop in pressure per degree centigrade amounted to 1/273 of the pressure at the temperature of melting ice, i.e., at 0°C. Absolute zero was thereby fixed at −273°C.

It is significant that Amontons visualized this absolute zero as a state of complete rest at which all motion would have ceased. This fact is important because it gives a hint about his ideas concerning the nature of heat. The meaning of heat and temperature and their proper measurement provided the scientists of the seventeenth and eighteenth centuries with some awkward problems. What, they asked themselves, is the nature of 'heat'? Clearly, the physical state of water, which had been heated over a flame, just before boiling was different from cold water. This could be shown by Amontons' thermometer. Also, something had gone into the water and this something had been provided by the flame. Alchemical tradition suggested that some fiery 'principle' from the flame had entered the water. On the other hand, a hot piece of iron dropped into water would also raise its temperature. Plausible explanations for all these observations could be provided by postulating the existence of a fluid, called caloric, whose only properties were heat and the ability to pass from one body into another on contact. Accordingly, caloric had been transmitted from the flame to the iron and subsequently from the iron to the water. Its concentration was measured by the thermometer and, as any fluid must, it flowed from higher to lower concentration. Eventually, iron and water had the same temperature, which means that they contained the same concentration of caloric. The great advantage of this concept of a material nature of heat was that it permitted quantitative statements. For instance, it is possible to measure out caloric in units. The amount of caloric which will raise the temperature of one gram of water by one degree is called a calorie. The trouble with caloric is that it is a weightless fluid since it was found that the piece of iron did not get heavier when it was heated. This made it difficult to fit it into the remainder of our physical world picture.

The idea of caloric as a weightless fluid was finally discarded early in the nineteenth century but the usefulness of the concept is demonstrated by the fact that it has been retained in less explicit form to this day. The concept of 'a quantity of heat' which is measured in joules, together with such other concepts as temperature, pressure and volume, is one of the basic notions of the discipline of thermodynamics.

The strength of thermodynamics lies in its use of these well-defined quantities which can be measured easily and unambiguously. The simple and straightforward formulae by which these quantities are linked with each other became the key to a host of baffling scientific and technological problems which often have been solved, thanks to the generalised nature of thermodynamics, with an almost uncanny ease.

The concepts of thermodynamics will be with us from now on for the rest of our story. They stand side by side with those of the atomistic interpretation of heat, the kinetic theory. The quantities with which thermodynamics deals are: temperature, measured on a thermometer; pressure, measured as the force exerted per unit area; and volume, measured by the size of a container. They all refer to observations on a scale many million times larger than that of an individual atom. Thus, in thermodynamics, the basic quantity of heat is the joule, which is the amount of heat required to increase the temperature of one gram of water by about a quarter of a degree on the centigrade scale (more precisely 0·239°). Nothing is said in this definition concerning the physical nature of heat. It can be a weightless fluid or something else.

Amontons' reference to absolute zero as a state of rest shows that, indeed, he thought of something other than the hypothetical substance 'caloric'. This something else is motion on the atomic scale. Two thousand years before Amontons, Democritus postulated that all matter is composed of tiny, indivisible building bricks, the atoms. From then on the atomic theory had its ups and downs but was never completely discarded. One of its attractions is that it permits an explanation of heat which does not need the weightless fluid but is based on something already known; the laws of mechanics. Newton and Amontons were contemporaries and it was only natural to apply Newton's laws which describe the motion of bodies in the skies and on earth also to those hypothetical small particles called atoms. Amontons does not seem to have gone beyond that vague reference but in 1738 the question was treated rigorously by the great Swiss mathematician and scientist, Daniel Bernoulli. In his famous treatise on 'hydraulics', he postulated that all 'elastic fluids', such as air, consist of small particles which are in constant irregular motion, continually colliding with each other and with the walls of the container. Since the collisions are perfectly elastic, the motion does not 'run down'. The particles behave somewhat like tennis balls with the difference that they don't get tired at all. Each of these atomic tennis balls when dropped on the floor will bounce back again and again, returning to the original height for ever and ever. The violence of the collisions, the speed of the individual particles, is registered by us as the sensation of heat. Bernoulli pointed out that his theory comes to the same result as was reached by Amontons experimentally in 1702.

There is great beauty in this kinetic interpretation of heat, not because it is particularly simple – and as we shall see it is not necessarily so – but because it explains the phenomena of heat completely in the well-known terms of mechanics. No new concept such as the weightless caloric is required.

It is the basic creed of science that the physical world has been created on a consistent and integral plan. We therefore expect that all our observations, however unconnected they may appear to us, must be part of this pattern. The discovery of this pattern is the sole object of all fundamental research. So far we are only at the beginning when we are still busy discovering new phenomena and thereby know that, even should we be extremely clever, we cannot hope to form adequate ideas of the pattern. It is like building up a jigsaw puzzle and at present we know that we haven't even got all the pieces. Neither do we know how large the puzzle may prove to be and there is an uncomfortable feeling that it may be of infinite size. Considering this state of affairs we must count ourselves lucky when here and there we find a few pieces which fit together, forming little islands of pattern which as yet bear no relation to other islands. Thus there is triumph indeed when we can link up two islands. The direct connection of the phenomena of heat with the laws of mechanics by means of the kinetic theory is such a triumph, one of the greatest in the history of science.

Nevertheless, it took another century for Bernoulli's kinetic theory to be accepted and we can easily see why. The explanation of heat as a sensation produced by an immense multitude of tiny pinpricks given to us by the speed of invisibly small particles is convincing. Moreover, as we shall see immediately, the relationships between volume, pressure and temperature discovered experimentally by Boyle and Gay-Lussac emerge as simple, straightforward consequences from the kinetic theory. What, however, is a quantity of heat expressed in terms of atomic motion? One can see straight away that here the concept of a weightless fluid which can be poured from one body into another is much more convenient. Moreover, as any billiard player knows, collisions between more than two balls are difficult to predict, so how can one tackle mathematically the collisions between millions and millions of fast moving particles? It was only when this last problem was solved satisfactorily that the kinetic theory of heat came into its own. In the meantime, as long as one doesn't ask for an explanation by means of a physical model, there is always thermodynamics. Its solutions may not be informative but they are always correct.

There are still today many problems in the phenomenology of heat which present no great difficulty for thermodynamic solution but which have remained quite intractable to kinetic interpretation. Whenever conditions are chosen in which we can regard the atoms or molecules as fully elastic tennis balls, kinetics yields a clear and simple picture, but once we have to take into account that these tennis balls exert forces upon each other, the difficulties become insuperable. On the other hand, the apparently formidable mathematical difficulty arising out of the multitude of individual collisions was solved by Maxwell and by Boltzmann in the second half of the nineteenth century. The mathematical method used is that of statistics which, as the name implies, had proved its worth in statecraft when dealing with human multitudes. Like most methods of statecraft, it disregards the individual and instead deals with averages. Just as the budget deals with average income and the insurance companies with average expectation of life, kinetic theory deals with average velocity. It does not hurt the insurance companies when some people die sooner than the average because others live longer than the average. At any moment in a gas there will be some atoms with much higher and others with much lower speed than the average, but this again does not matter as long as the statistical treatment is extended over very large numbers.

Let us now apply these considerations to the simple case of a space filled with a gas, say, air, of which we can change the volume, for instance a cylinder with a piston. First of all we keep the piston stationary in one position and we measure the pressure of the gas, i.e., the force which the gas molecules exert on the piston. This can be done by finding out what weight has to be put on to the piston in order to keep it in place. The force is given by the number of molecules hitting unit area of the piston in unit time and on the average speed of the molecules. The latter is a measure of the temperature and since we want to carry out our first experiment without a change in temperature this average speed will remain the same. The first experiment consists in putting so much more weight on to the piston that the gas is compressed to half its original volume. We then find with Boyle, who did the same thing three hundred years ago, that exactly twice as much weight is needed. This result accords well with the theory since the same number of molecules are now confined to half the volume and therefore hit the piston twice as often.

For our second experiment we raise the temperature but do not change the position of the piston. Now the speed of the molecules is increased and with it the force of each individual impact. In addition, the greater speed will also cause the piston to be struck more frequently. The increased force on the piston is therefore due to the square of the molecular speed. This rise in the pressure is, as was observed by Amontons, directly proportional to the rise in temperature recorded by his thermometer. In this way we have now found the meaning of temperature in our atomistic model. It is the squared velocity of the molecules; which is the same as their kinetic energy.

Finally, we investigate what happens in the process while we push down the piston by increasing the weight and now we shall take no steps to keep the temperature constant. Let us look for a moment at one particular molecule which is about to strike the piston. It has a certain velocity before the collision takes place. However, when it bounces back its speed is now greater because the piston was not stationary but was moving with its own velocity against the molecule. The returning molecule will at its next collision transmit this higher speed to another one, and so forth. In this way the average velocity of all the molecules in the cylinder is increased and such an increase corresponds, as we have just seen, to an increase in temperature. In other words, a gas which is compressed

1·1 Cailletet's cooling device may be considered as part of a single stroke of an expansion engine.

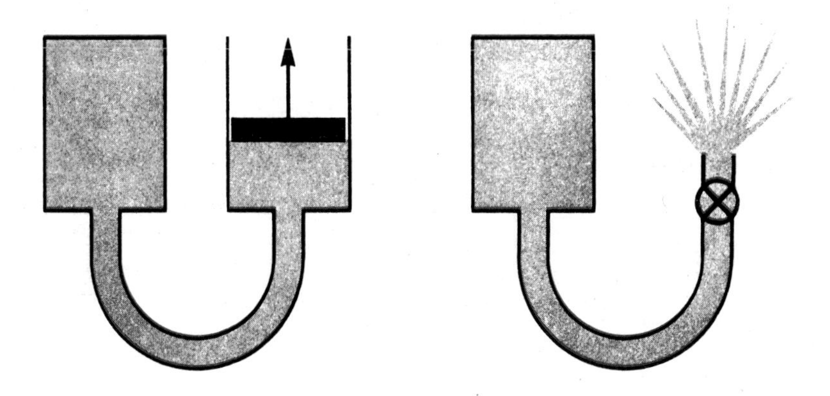

heats up. This is an effect well known to anyone who has pumped up a bicycle tyre and has noticed that the pump gets hot.

The opposite is true as well. If we decrease the weight on the piston and allow the gas to push it out, the molecules strike a receding piston and come back with lower speed. The gas cools on expansion. We have now arrived at the explanation of Cailletet's experiment. He used no piston but one can easily understand what happened in his case by dividing a cylinder into two parts which are connected by a tube and only one of which has a piston (figure 1·1). The other shall be Cailletet's strong walled glass tube, containing compressed oxygen gas. If now this gas is allowed to expand by pushing back the piston in the right-hand cylinder, the oxygen in both the right *and* the left-hand container will cool down since, thanks to their collisions with each other, the decrease in speed is shared by all molecules. As the piston recedes it is followed by a rush of gas through the connecting pipe. As a matter of fact, the molecules in Cailletet's tube on the left are quite unaware of the existence of the piston and, instead of going to all the trouble of providing a cylinder and piston, it is much simpler and cheaper to have a tap at the end of the connecting pipe. When this is opened the compressed oxygen rushes out and what remains in the strong-

1·2 A continuous cooling cycle employing a compressor A, an expansion engine B and a heat exchanger D. The heat abstracted at B is removed from the cycle by the cooling water in C.

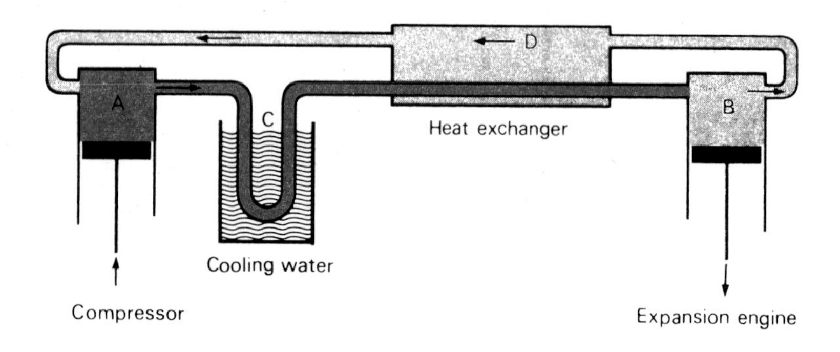

walled tube cools down so much that droplets of liquid are momentarily formed. The tap at the end of our pipe is, of course, nothing but the accidental leak in Cailletet's apparatus which led him to success.

Cailletet had tried to liquefy acetylene by subjecting it to high pressure and it seems fairly certain that he did not immediately realise the full meaning of his successful mishap. We have just seen how his experiment can be explained in terms of the kinetic energy but there is, of course, also an explanation in the language of thermodynamics. It is older than the kinetic one and lends itself better to quantitative calculation but it is less informative and never tells us what happens in detail in the expanding gas. The foundation of the thermodynamical treatment was laid half a century before Cailletet by the French engineer officer Sadi Carnot in his theory of the heat engine. In his treatise 'Réflexions sur la Puissance Motrice du Feu' which was published in 1824 and which was to form the basis of power production as well as of the science of thermodynamics, he relates the mechanical energy required to compress a gas with the increase of temperature produced by it. Conversely, the energy supplied by the gas in pushing out the piston must result in cooling. In these processes one thus deals with a transformation of energy from mechanical work into heat, and vice versa. Applying these considerations one can construct a simple cooling device which consists of two reciprocating piston engines, one (*A*) for compressing the gas, and the other (*B*) for letting it do mechanical work by expanding (figure 1·2). The compressed gas leaving *A* is hot and this heat of compression is removed by cooling water in the trough *C*. The gas now passes on to the expansion engine *B* where heat energy is transformed into mechanical work. The gas which has thereby been cooled leaves *B* and, before entering again *A*, flows at *D* through a tube which surrounds the pipe carrying the compressed gas into *B*. The object of this device *D* is to cool the incoming compressed gas by the counterflow of outgoing cold gas from the expansion engine. This 'heat exchanger' sees to it that after each stroke of *B* the gas coming into *B* for the next stroke will be colder than before. The whole arrangement is a closed cycle in which heat is constantly removed by the cooling water and where *B* gets progressively colder. This can go on until finally a temperature is reached where the gas in the cycle becomes so cold that it begins to liquefy in *B*.

A cooling apparatus on these lines was, in fact, patented by Siemens as early as 1857 but when, twenty years later, Cailletet made his experiments, no reference was made by anyone to the patent. This suggests that at the time the connexion between Cailletet's method and a piston-type cooling engine was not clearly recognised. A number of Siemens engines were, in fact, built, and in 1862 the Scottish engineer Kirk produced a most successful model for the cooling of shale oil in a refinery process which, incidentally, was capable of freezing mercury, i.e., reaching a temperature of at least −39°C. However, the people developing these machines were not scientists but practical engineers whose aim was to provide refrigeration in ships so that meat could be brought in good condition from Australia to England, which was done for the first time in 1879.

The first attempt to liquefy air with an expansion machine was made by Solvay, probably in 1887 when a German patent was granted to him. This describes three possibilities, two of which are expansion engines. One was tried but achieved only about half the required temperature drop. Since the original notes of the experiment are lost, it is not known which of his three schemes had been tried.

Success was eventually achieved by the French engineer Georges Claude whose systematic and persistent efforts led to air liquefaction in 1902. The technical difficulties which had to be overcome were considerable. First of all, the expansion engine had to be thermally well insulated since its ultimate operating temperature was below −150°C and heat influx had to be reduced to a minimum. Secondly, lubrication of moving parts is made exceedingly difficult since, at these low temperatures, oil will freeze. Petrol ether which remains liquid down to −140°C can be used, but better results were achieved in the end by employing a dry leather packing. A small amount of air which leaks out between the piston packing and the cylinder wall is the real lubricant in this case. Claude would probably not have persisted in the face of all these difficulties for the sake of a scientific experiment but at the turn of the century air liquefaction had assumed the aspect of a major industrial undertaking for the separation of oxygen from the atmosphere. He was guided by the success which his German and British competitors had achieved with a different type of process seven years earlier.

Temperatures of well below −200°C were reached by Claude's expansion engine in 1920, using hydrogen gas instead of air as working substance. Here again the object was gas separation, hydrogen from coke oven gas, and not the liquefaction of hydrogen itself. Nobody seems to have seriously considered the expansion engine as a tool for scientific low temperature research. When in 1933 I visited the Russian physicist Peter Kapitza in his new laboratory in Cambridge, I noticed some metal pressings lying about which I took to be parts of a refrigeration device. Kapitza agreed but avoided being specific. He said that he preferred to see it work before talking about it. A year later he announced his success. He had built a helium liquefier which contained an expansion engine operating only ten degrees above absolute zero, an ingeniously designed and superbly executed engineering achievement. Finally, in 1946, a helium liquefier on similar lines was designed and built by Professor Collins of the Massachusetts Institute of Technology and marketed commercially by the A. D. Little Company. Well over a hundred of these machines have been sold all over the world and have caused a revolution in low temperature research. Liquid helium which until then had been a scarce commodity, available only in a few laboratories, suddenly became accessible to a large number of research centres, both academic and industrial. The result has been a tremendous increase in scientific and technological work at very low temperatures so that the period until 1946 is now often referred to as B.C., before Collins.

The piston engine is not the only device by which heat can be transformed into mechanical work. For large-scale power production, such as in generating stations, ocean liners and now also in aircraft, it has been replaced by the turbine. There are reasons for the greater thermodynamic efficiency of a turbine which, however, are outside the scope of our story. The same reasons make the turbine even more attractive for the abstraction of heat, that is, for the productions of cold. Patents for turbines to be used in gas separation were taken out by Thrupp in England and Johnson in America in the late 'nineties. Liquefaction turbines were first successfully operated in the early 1930s by the Linde Company in Germany but, for reasons of industrial or military secrecy, little information became available in the subsequent years. It was again the engineering genius of Kapitza, now in Moscow, who in 1939 presented a detailed analysis of the superiority of the turbine as a liquefaction device. Besides its fundamental advantages based on thermodynamic considerations, the turbine operates at much lower pressure than the piston engine, making the whole plant much safer and cheaper in construction. Today the turbine has become standard equipment for the mammoth low temperature installations feeding steel works and rocket bases. The fog seen by Cailletet in 1877 has led to a technological development of gigantic size and with ever-increasing ramifications.

Let us now return to the missing link between Cailletet's experiment and the expansion engine. The expansion engine yields cold because it does mechanical work, but where is such work done in Cailletet's apparatus? All that happens is that the compressed oxygen gas is released into the air. The clue is provided in figure 1·1 where we have replaced the right-hand side of the cylinder by a stop cock. Once this tap is opened the oxygen does a little more than just rush out. It lifts up the earth's atmosphere by a very small amount and it is this work which produces the cooling. In other words, Cailletet's experiment represents nothing more than an incomplete single cycle of the cooling device shown in figure 1·2. The oxygen gas is first compressed at *A* to 300 atm, it is pre-cooled to −29°C by evaporating sulphur dioxide in *C*, and then fed, without using a heat exchanger, into *B*. *B* in Cailletet's case corresponds only to part of an expansion cylinder (the left-hand side in figure 1·1) and, when the tap is opened, only part of a single expansion stroke is performed. A small fraction of the expanding gas remains in the glass tube where the droplets of liquid oxygen make their appearance. The majority of the droplets have escaped through the valve and evaporated as quickly as they were formed.

While Cailletet's experiment demonstrated for the first time the earth's atmosphere in the liquid state, it is hopelessly inefficient as a method for gas liquefaction. When, a few years later, better ways of making liquid air were found, Cailletet's method was discarded and remained nothing better than a historic curiosity until 1932. In that year it was revived in its old unchanged form by F. E. Simon, who demonstrated that in the special case of helium it is, while still fairly inefficient, a simple and convenient method for obtaining the liquid.

Cailletet followed up his success by improving his apparatus. In 1882 he cooled the glass capillary with liquid ethylene instead of sulphur dioxide, reducing the starting temperature of his experiment from −29°C to −105°C, and on expansion he observed violent turbulence in the tube but the spray of liquid oxygen again evaporated immediately. The next decisive step in the mastery of low temperatures had again eluded him. This next step had been clearly formulated at the memorable meeting of the Academy on Christmas eve 1877 by another great pioneer of nineteenth-century physics, Jamin. Jamin said: 'The possibility of liquefying oxygen has now been proved. . . . The decisive experiment has still to be made. It will consist in keeping liquid oxygen at the temperature of its boiling point.' He made a clear distinction between the momentary observation of a mist of droplets and a liquid with a meniscus quietly boiling in a test tube.

It was not granted to Cailletet and the French physicists to achieve this. He carried on his work for many years to come and when, in 1913, he died in Paris at the age of eighty he had the satisfaction of seeing that his original discovery had turned into a new branch of physics. All the gases had been liquefied and absolute zero had been approached to within one degree. The door had been opened to a strange new world of baffling physical phenomena such as the complete disappearance of electrical resistivity. His old laboratory in Chatillon-sur-Seine is still there, a room full of his apparatus, overlooking his vegetable garden, and lovingly cared for by his descendants.

2 Cracow 1883

The impact of Cailletet's observation on the development of scientific work is marked by the fact that in 1882 a Polish scientist then working in Paris made reference to the great number of Cailletet apparatuses which were manufactured there by the famous instrument-making firm of E. Ducretet. He pointed out that Cailletet kept nothing of the experimental details secret from his colleagues, many of whom were keen to repeat and extend his work. The Pole was one of them and he bought an apparatus from Ducretet to take it back to Cracow where he had just been appointed to the chair of physics. His name was Szygmunt Florenty von Wroblewski. The old Jagiellonian University of Cracow belonged at that time to the Habsburg monarchy, which explains the German predicate of nobility in Wroblewski's otherwise Polish name. On arrival in Cracow Wroblewski found in the Chemistry Department a man of his own age – he was in his middle thirties – and with the same interest in gas liquefaction. This was K. Olszewski, who for the past thirteen years had struggled along with old and insufficient equipment. He, of course, was delighted with the arrival not only of Wroblewski but also of the modern liquefaction apparatus. The two of them set to work in February 1883, and on the 9th April they succeeded where Cailletet and the others had failed: liquid oxygen was quietly boiling in a test tube of their apparatus.

To have achieved this in two months or less is almost incredible. It was a feat which will be envied by many modern researchers who are armed with ample funds, equipment and assistance. In order to explain it at all, we must have regard to a number of factors. Wroblewski was well acquainted with the Paris experiments at which he had been present. Olszewski, who for more than ten years had wrestled with antiquated and makeshift high-pressure equipment, had acquired practical experience second to none. Cailletet's apparatus of the year before required only small modification to make it yield liquid oxygen in a stable state. In fact, Cailletet had almost succeeded and all that was needed was a new idea. For this, however, the Poles can take full credit because they succeeded

2·1 The apparatus in which Wroblewski and Olszewski first produced liquid oxygen 'boiling quietly in a test tube'.

by a better understanding of the physical principles involved.

Of the two modifications which they introduced, the first is trivial. They bent over the glass capillary so that any liquid oxygen collecting should not run out through the expansion tap but would be caught in the bottom of the capillary (figure 2·1). The second change was the significant one. Like Cailletet, they used liquid ethylene to cool the capillary but, instead of allowing it to boil under atmospheric pressure, they pumped off the vapour above the liquid, down to a pressure of 2·5 cm mercury column, i.e., to one-thirtieth of an atmosphere. In this way the temperature was reduced to about −130°C, and when oxygen gas under high pressure was introduced into the capillary, they saw little droplets forming on the glass wall and collecting as liquid at the bottom of the capillary. They had thus liquefied oxygen without making use of Cailletet's original device of expanding the gas.

In order to understand the underlying principle of this type of liquefaction we shall have to go back to earlier experiments which

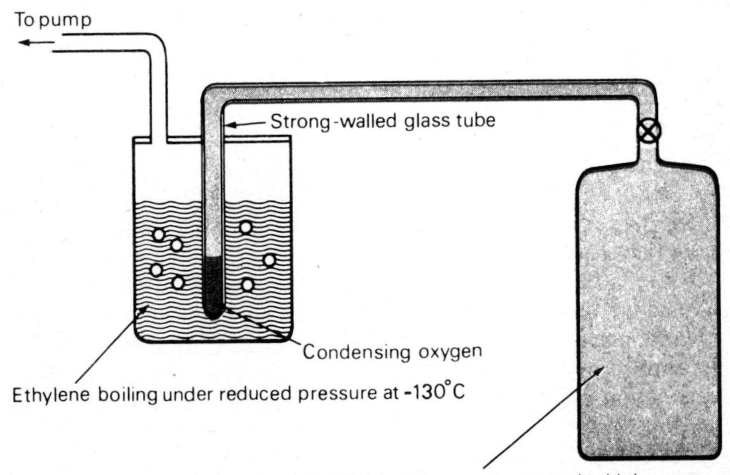

had elucidated the role played by both pressure and temperature in the liquefaction process and the nature of the equilibrium between liquid vapour and gas. Before we do this, a few more words must be said about the development of the Cracow low temperature school.

The first note on the decisive experiment which was submitted to the meeting of the French Academy on the 16th April 1883 appeared under the names of both experimenters and so does the full paper in *Annalen der Physik und Chemie*. We thus have no indication which of the two had the idea of pumping off the ethylene vapour, but it may be significant that the names are not in alphabetic order, Wroblewski's being given as the first author. Textbooks usually refer to Olszewski and Wroblewski as the leaders of the Cracow School, but, in fact, the collaboration did not outlast another six months. In October 1883 we find Olszewski back at the Chemistry Department doing his own low temperature experiments, while Wroblewski continued in much the same work at the Physics Laboratory. What exactly happened cannot easily be found out at this stage but many years later Olszewski refers pointedly to the 'dissolution of collaboration'. It seems that, in spite of working on exactly the same subject in the same university, they remained poles apart until Wroblewski's untimely death five years later. Working late in the laboratory he upset a kerosene lamp on his desk and perished in the flames. Today his scorched note papers are exhibited in the ancient hall of the Jagiellonian University where they rest not far from the entry signed by a famous student more than four hundred years earlier; one Nikolaus Kopernick.

In our account of the expansion engine given in the preceding chapter we have come across the kinetic and the thermodynamic explanations of the cooling process, but nothing was said about the mechanism of liquefaction. Lavoisier's prophetic statement on the liquefaction of air which was quoted in the Academy at their meeting on Christmas eve 1877 visualised cold as the important agent. However, even in Lavoisier's time it began to be realised that pressure, too, played some part in turning a gas into a liquid.

It was all complex and confusing but, on the other hand, Robert Boyle's investigation into 'the spring of the air' had revealed a beautifully simple relationship between the pressure and volume of a gas. At the end of the eighteenth century M. van Marum of Haarlem was engaged in experiments designed to see whether Boyle's law was true for all gases or only held in the case of air. One of the substances which he chose for his investigation was ammonia gas and it was here that he made an important discovery. As he used increasingly higher pressures the gas volume failed to decrease proportionally, as Boyle's law would predict. Finally, when he had reached about 7 atmospheres, the pressure suddenly did not rise on further compression but the volume continued to decrease. What was happening was that, instead of being further compressed, the ammonia gas at this pressure turned into liquid ammonia. From then on, all that further reduction of the volume could achieve was an increase in the amount of liquid in the cylinder at the expense of the gas volume. Liquefaction of ammonia had been accomplished by pressure only, and without any necessity to lower the temperature.

Van Marum's observation was followed by countless attempts at turning known gases by compression into the liquid state. Some of them were indeed liquefied but others remained gases even under the highest available pressures. One of these experiments had an important bearing on the later history of gas liquefaction, although it had not been designed primarily for this purpose. At the Royal Institution in London Humphrey Davy, who, among other notable achievements, is also remembered for his invention of the miner's safety lamp, was carrying out investigations into the properties of chlorine compounds. Sometime in 1823, Michael Faraday, who was then Davy's laboratory assistant, heated in a sealed glass tube one of these compounds in order to study its chemical decomposition. He and a friend of Davy's, Dr. Paris, who was watching the experiment, were mystified by some droplets of an oily liquid which appeared at the cold end of the tube (figure 2·2). Dr. Paris left, wondering whether the materials used had been impure, but

2·2 Faraday's liquefaction of chlorine.

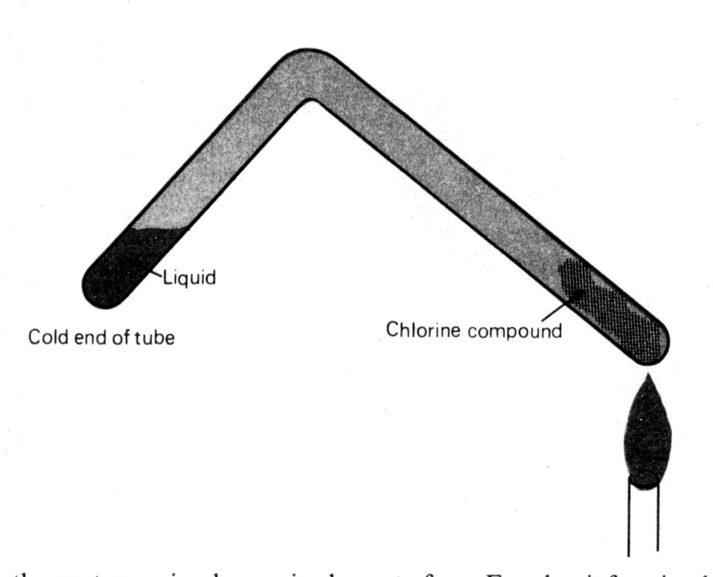

the next morning he received a note from Faraday informing him that the droplets had been liquefied chlorine gas. Faraday recognised that, in addition to the pressure which he had created in the sealed tube by heating, temperature played a part since the liquid had collected at the cold end. When he returned to these experiments in 1826 and again in 1845, now as Director of the Royal Institution, he varied them by immersing the unheated end of the tube in a cooling mixture instead of leaving it at room temperature. This stratagem proved highly successful and he was able to liquefy a number of gases which had resisted earlier attempts. However, oxygen, nitrogen and hydrogen showed no sign of liquefaction and many scientists believed that these were what they called 'permanent gases'.

The idea of permanent gases had received support through a series of negative, if spectacular, experiments. If air did not become a liquid at the highest pressure which could be produced in an experiment, a still higher pressure had to be created. Ingenious devices were tried to achieve this. Aimé compressed oxygen and nitrogen to well over 200 atmospheres by lowering them in suitably designed cylinders to depths of over a mile into the ocean. An otherwise undistinguished medical man in Vienna, Natterer, became the outstanding pioneer in the construction of high pressure compressors. In 1844 he also arranged for the manufacture of these machines by the 'well renowned mechanic Herr Kraft', 'in order to save the friends of physics and chemistry the discovery of all hindrance and the invention of the most advantageous, most suitable and safest design of the whole apparatus'. The machines cost 100 florins and were tested to 200 atmospheres. Natterer himself was going to be well ahead of his customers, stating that he was 'of a will to continue compression to 2,000 atmospheres'. In fact, he exceeded his aim, reaching a few years later something of the order of 3,000 atmospheres, a quite astonishing engineering achievement in the new field of high pressure technique. His record was not beaten for a long time to come, and forty years later we find Olszewski relying for his successful experiments on an old Natterer compressor which he had discovered in the otherwise poorly-equipped Cracow laboratory when he arrived there as a research student. Natterer himself had shown one thing at least; that even at a pressure of 3,000 atmospheres air did not liquefy. Oxygen and nitrogen seemed truly permanent gases.

Faraday was not the first to realize that both pressure *and* temperature play their part in the change from a gas to a liquid. In 1822 Charles Cagniard de la Tour, an Attaché in the Ministry of the Interior in Paris, resolved to find out what would happen to a liquid enclosed in a sealed space and then heated up. He first chose alcohol as a suitable substance for investigation and enclosed it in what was essentially a pressure cooker fashioned from the closed end of a strong walled gun barrel. This, he felt, would safely withstand both pressure and temperature but, of course, he could not see what was happening inside. So he decided to listen. He enclosed a quartz ball with the alcohol and found that it made a different noise rolling about in the liquid than it made in air.

He then closed the vessel and heated it up, noting the noise all the time. He found that eventually, at a sufficiently high temperature, the alcohol had completely changed into the gas phase and that there was no liquid left in the vessel. Since he now wanted to see what happened and how it happened, he made his next observations with sealed-off glass tubes which he heated and in which he had enclosed increasing fractions of liquid. Although he used strong-walled tubes, they usually exploded when as much as half the volume was originally occupied by liquid. Even so, in the end he felt justified to conclude that at pressures above 119 atmospheres alcohol could not exist in the liquid form and that when the change took place the liquid in his tube turned suddenly before his eyes into a gas.

On the strength of this conclusion we must regard Cagniard de la Tour as the discoverer of the salient feature of the gas-liquid equilibrium, the so-called critical point. However, the nature of this equilibrium remained far from clear and its proper elucidation was only gained by a superb series of experiments carried out by Thomas Andrews at Queen's College, Belfast, between 1861 and 1869. Andrews considered his work as the continuation of de la Tour's investigations. However, he now had at his disposal much better apparatus than his French predecessor, and he also drew a lesson from the latter's failures. Cagniard de la Tour had been moderately successful with alcohol but water, with its higher boiling point, had required the application of pressures which were too high for his glass tubes to withstand. Andrews therefore chose for his work carbon dioxide (CO_2), a substance which is gaseous at normal temperatures and, as he expected, the pressures required for studying the whole range where gas and liquid are in equilibrium were relatively low. The kind of measurements which he performed were of the same type as those undertaken by van Marum on ammonia. He determined, at different temperatures, the change in the volume of a given quantity of the substance when the pressure is varied. The resultant curves are called isotherms because they each refer to one and the same temperature.

2·3 Isotherms of an 'ideal' gas for four different temperatures. According to Boyle's law the product of pressure and volume must be the same for any values of P and V at constant temperature. The grey and pink areas are of equal size.

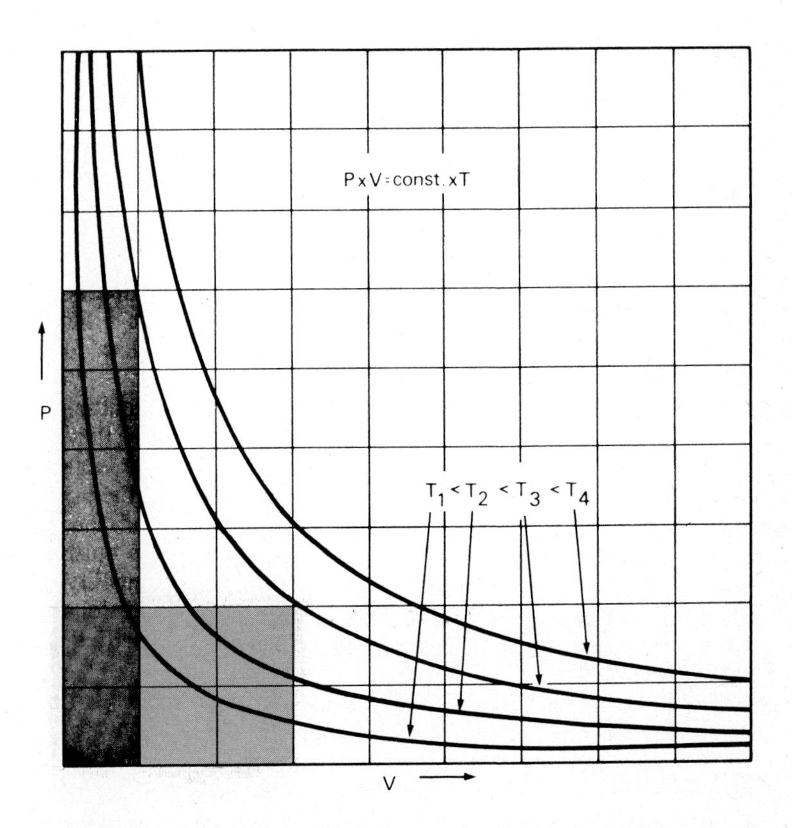

These are the same measurements as had been made by Boyle and by Mariotte, on air at room temperature, two hundred years earlier. We have already referred to the relation between pressure P and the volume V which they had discovered and which can be written in the form of the simple equation $P \times V =$ constant. Thus, if air is to be confined to half its original volume, twice the original pressure is required, for a third of the volume three times the pressure, etc. We have also seen that this simple relationship can be accounted for by the kinetic theory. Varying now the temperature and again measuring isotherms, one finds the result first noted by Amontons, later again by Charles and by Gay-Lussac, which can be written down equally simply as $PV = \text{const.} \times T$, where T is the temperature measured from absolute zero upwards. Making a diagram in which we plot P and V as ordinate and abscissa, and marking all points for which the product PV has the same value, we obtain a symmetrical curve which is the isotherm of air for a given temperature T (figure 2·3). For a higher temperature we get a curve of similar shape, but, since PV now has a larger value, it lies above the first isotherm. Equally, an isotherm for a lower temperature will lie lower.

These isotherms are all smooth curves and the mathematicians call them rectangular hyperbolae. What interests us is that they do not show any irregularity which we might expect if the air were to change somewhere in our diagram from the gaseous into the liquid state. Nevertheless, Cailletet was able to liquefy air and the inescapable conclusion is that in some respect our diagram must be wrong. Putting it into different words, we must expect that Boyle's law is not universally valid and that, under certain conditions, it will break down. Andrews' aim was to discover what exactly these conditions are.

An indication of what he might expect had been provided by van Marum's observations, and he, after all, had been interested in precisely the same problem; to see whether Boyle's law was valid for ammonia. Van Marum had been going along one of these isotherms (that at room temperature) and had found that when he raised the pressure, the volume first decreased more than, according to Boyle's law, it should have done. Finally, he reached the pressure where further decrease of the volume did not result in any additional compression of the ammonia gas but, instead, produced the appearance of liquid ammonia. Andrews found that van Marum had been quite correct in his observations but he now extended the scope of the investigation by measuring a great number of different isotherms. The set of curves which he obtained look quite different (figure 2·4) from those of Boyle's law. All those corresponding to low temperatures show the behaviour already found by van Marum. They all have a flat portion, corresponding to the region where the liquid condenses from the gas. Following any of these isotherms from large to smaller volume, i.e., starting on the right-hand side, we encounter the rise and then a kink where the level portion starts. Here the very first droplets of liquid appear. When now the volume is further decreased, more and more of the gas turns into liquid until, at the end of the level stretch, there is no gas left at all. From now on any further increase in pressure hardly changes the volume at all, showing that the liquid phase is highly incompressible.

The flat part of the isotherm reveals an important fact. Since the pressure remains constant while more and more of the gas condenses into liquid, the pressure of the gas in contact with the liquid must be always the same, quite independent of whether a small or a large fraction of the volume is occupied by liquid. It also is apparent from Andrews' diagram that this equilibrium pressure rises as we go to higher isotherms, i.e., as the temperature is increased. Moreover, we also notice that the flat part becomes shorter until a singularly important isotherm is reached which has no true flat portion at all but just one point at which the direction of the curve changes its sign. The higher isotherms are now all ascending smoothly over the whole range of pressure and volume, and if one goes to still higher temperatures, the isotherms attain more and more the shape of true rectangular hyperbolae. This is then the region in which Boyle's law is valid.

Andrews' results not only yielded a wealth of new facts but they also presented a beautifully complete and satisfying picture of the relation between the gaseous and liquid states of aggregation. Moreover, all the puzzling and contradictory results obtained in the past now fell into place, each suddenly making sense in the newly-revealed pattern. Most important of all, Andrews' careful measurements opened the way to an understanding of the strong forces of cohesion which are vested in each atom but never reach

2·4 Isotherms of a real gas (CO_2) as measured by Andrews. They approximate Boyle's law only at high temperatures (T_7). At low temperatures they are more complicated and below the critical point there is a region of liquefaction.

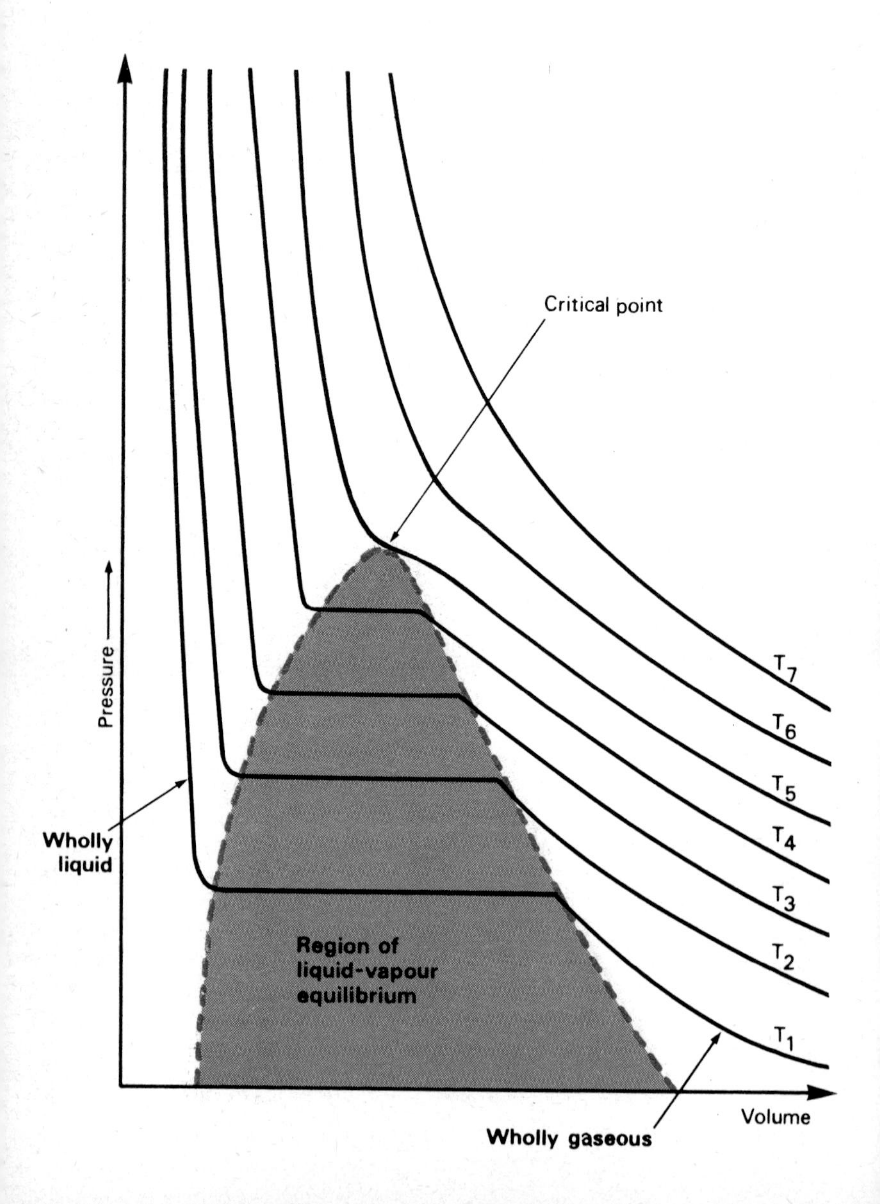

the dimension of ordinary macroscopic observation. It should also be noted that, while Andrews' observations were confined to carbon dioxide, the pattern is quite generally valid. The actual temperatures, pressures and volumes vary from one substance to the other but the variation of the isotherms with temperature is much the same whether we take water, hydrogen or iron.

In the lower part of Andrews' diagram one can clearly mark out the regions in which the substance is either wholly gaseous or wholly liquid or, as in the flat portion, partly in the liquid and partly in the gaseous state. This region where these two phases are in equilibrium, that is, where we can see a meniscus separating liquid and gas, becomes, however, narrower as one moves to higher temperatures. Finally, having arrived at our singularly important isotherm, it disappears completely. At any higher temperature the flat portion is missing and there is no region in the isotherm where we can see a meniscus. As Andrews himself said of this condition: 'If anyone asks whether it is now in the gaseous or in the liquid state, the question does not, I believe, admit of a positive reply.' On the other hand, Andrews' diagram leaves no doubt where a meniscus can and where it cannot possibly appear. Any isotherm with a flat portion indicates liquid and gas in equilibrium. At the point where the singularly important one changes direction, the last vestige of this equilibrium disappears. This is the 'critical point' of the substance. Apart from a critical pressure and a critical volume, it marks, more important than either, the 'critical temperature'.

The critical temperature of carbon dioxide was found by Andrews to be 31°C, and it is clear from his result that above this temperature even the highest pressure can never yield a state with a meniscus. This explains at once why Natterer, in spite of the enormous pressure which he was able to apply, failed to liquefy oxygen. As Wroblewski showed later, the critical temperature of oxygen is −118°C, and the critical pressure 50 atmospheres. It also now becomes clear why Cailletet's second attempt in 1882 to get liquid oxygen in stable form was unsuccessful, and why the Poles

2·5 The pressure–temperature diagram of state gives the curves at which the solid, liquid and gaseous phases of the same substance are in equilibrium. It should be noted that the liquid and gaseous phases are not completely separated but merge into each other above the critical point.

reached their goal. Cailletet's coolant, liquid ethylene, boiling at atmospheric pressure provided him with a starting temperature of −105°C, thirteen degrees above the critical point, where even Natterer's 3,000 atmospheres would not have yielded liquefaction. In the Cracow experiments, liquid ethylene was boiling under reduced pressure at −130°C, well below the critical point, and a pressure of only about 25 atmospheres was required to produce steady condensation of the liquid. The Poles had reached an isotherm with a flat portion.

We have used Andrews' diagram not only for its historical interest but also because it illustrates in a clear and convincing manner the significance and the boundaries of the liquid state, and we shall have to return to it later. However, for many purposes another form of representation is both more convenient and more informative and this, too, we shall have to use frequently. In it we plot not pressure and volume but pressure and temperature as ordinate and abscissa.

We can also take this opportunity of acquiring a habit which will make life simpler when we deal with low temperatures. This is to free ourselves from any artificial starting point of the conventional temperature scales which, if one follows Celsius, is at the freezing point of water, while for those who believe in Fahrenheit, it is enshrined in a mixture of ice and salt. Since Amontons and Gay–Lussac provided us with an absolute zero at −273°C, we might as well start counting from there, if for no other reason than that we only have to count in one direction – upwards. Nevertheless, one still has to fall back on convention in choosing the size of the degree in which we count, and here, by an admittedly arbitrary decision, the degree of the centigrade scale was adopted. Since °F and °C have been used as symbols for the two conventional scales, °K became the designation for the absolute scale, in honour of Lord Kelvin, who removed from the concept of temperature the limitations due to a thermometer, and based it on pure thermodynamics. In the SI system, the unit of temperature is still the degree Kelvin, now called simply the kelvin, and designated K instead of °K.

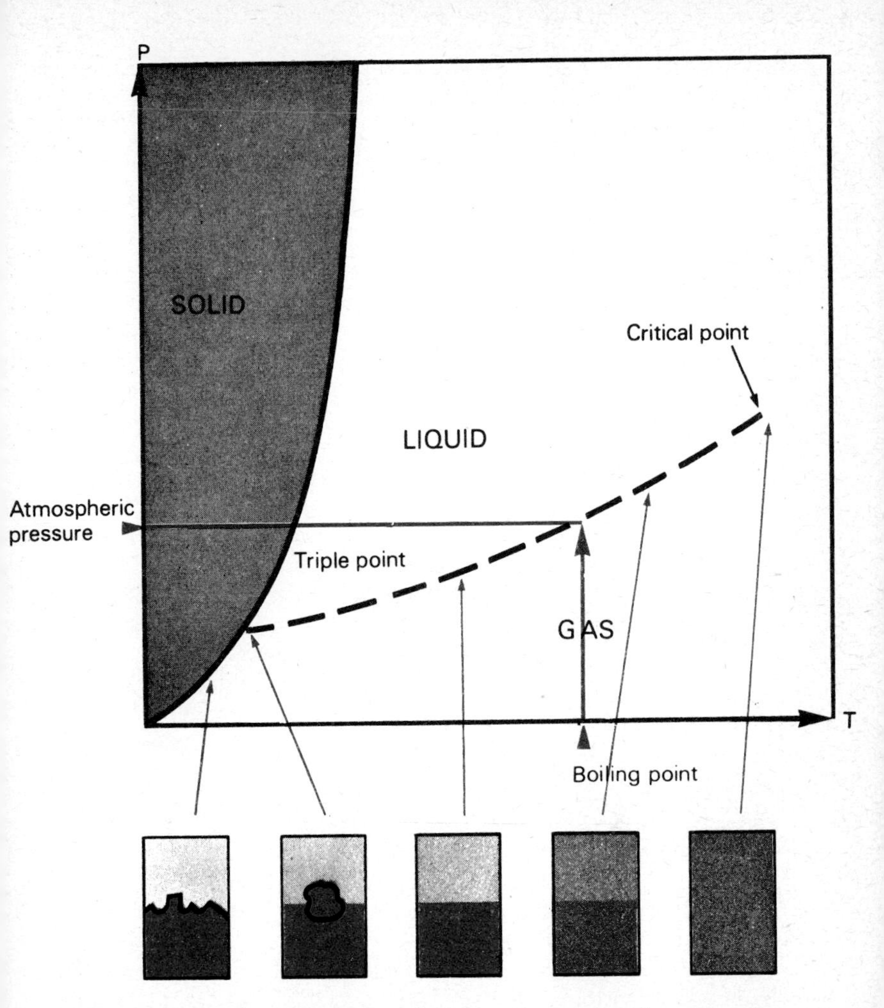

Zero is at the absolute zero and the temperature of melting ice (0°C) is at 273 K. Water then boils at 373 K, which is, of course, +100°C. This dispenses with tedious minus signs and subtractions, and henceforth we shall regard the critical temperature of oxygen not as −118°C but as (273−118=) 155 K.

In our pressure–temperature diagram (figure 2·5) the lines denote the boundaries of the different states of aggregation. The full line encloses the solid state. It merges with the temperature axis at zero pressure at absolute zero. From there on it rises, at first slowly and then more and more rapidly to the top of any diagram which we can draw today. This means that at the highest pressures which we can achieve in the laboratory there is always a definite temperature at which a substance will melt. What happens to this 'melting curve' beyond the limits of experimental investigation we cannot say. It is one of the unsolved problems of physics which we must leave, however much it may intrigue us, because our search leads us in the opposite direction. At sufficiently low temperatures the solid state always borders on the gaseous state and not, as one might easily have thought, on the liquid state.

We can lend some reality to this diagram by considering the case of water and see what will happen to a vessel filled with it as we change the temperature. In order to avoid the complication provided by the atmospheric air, we first evacuate the vessel and we also make it strong enough to withstand pressure without bursting. At some low temperature, well below freezing, we see nothing in it except ice. A pressure gauge attached to the vessel will record a small pressure which shows that the space above the ice is not empty. It is, in fact, filled with water in the gaseous phase, that is, with water vapour, and the pressure of this vapour has the value indicated by the full curve in our equilibrium diagram. We are in the region where the solid and the gaseous state are in equilibrium. As we warm up our vessel the pressure increases gradually, showing that some ice is evaporating. At last the temperature is reached where the ice begins to melt and there are now ice and water visible in the vessel. In addition, there is the vapour which we cannot see. This temperature is called the triple point because all three phases, solid, liquid and gas, are here in equilibrium. At any lower temperature there will be no water and at any higher temperature no ice.

Warming up the vessel beyond the triple point, we see only water but we also notice that the water level falls with increasing temperature. The inescapable conclusion is that some of the water evaporates and this is, indeed, proved by the fact that the pressure of the vapour increases steadily. We are now moving along the broken curve and its gradual rise corresponds to the rise of the flat portions of the isotherms in figure 2·4. Finally, something most extraordinary happens. Quite suddenly, at a certain temperature, the meniscus separating water from its vapour disappears from sight. Lowering the temperature again by a small fraction, the water level reappears at exactly the same position where it so dramatically vanished. We have reached the critical point.

Strange as the disappearance of the meniscus may seem, it is not difficult to explain. As the vessel was warmed up, the water in it was becoming less dense, as with any substance that is heated. On the other hand, more and more water evaporated, which means that the vapour became increasingly denser until finally the temperature was reached at which the density of the vapour had attained the same value as that of the liquid. At that moment all distinction between the liquid and the gaseous phases vanish, they have become identical. The state now reached is that described by Andrews when he says that the question whether the substance is a liquid or a gas has no answer. There is thus no point in talking about gas liquefaction unless a temperature can be reached in which a meniscus is formed between the two phases. Our diagram shows that no such meniscus can ever be formed at a temperature which is higher than the critical.

The pressure–temperature diagram allows us to determine at a glance the possible states in which a substance can be found at any given temperature. Before we use it in the discussion of gas liquefaction an explanation is required why the familiar term 'boiling point' has so far been omitted in our considerations. The reason is that, while of considerable practical importance, the boiling point has no fundamental significance and only exists when above the liquid there is also air. It is merely the temperature at which the vapour pressure of the substance becomes equal to the pressure of the atmosphere. Since the latter varies with elevation, water boils in Mexico City at 93°C instead of 100°C in London, and the boiling points of liquid oxygen are 87 K and 90 K respectively. The triple points and the critical points of water or oxygen, on the other hand,

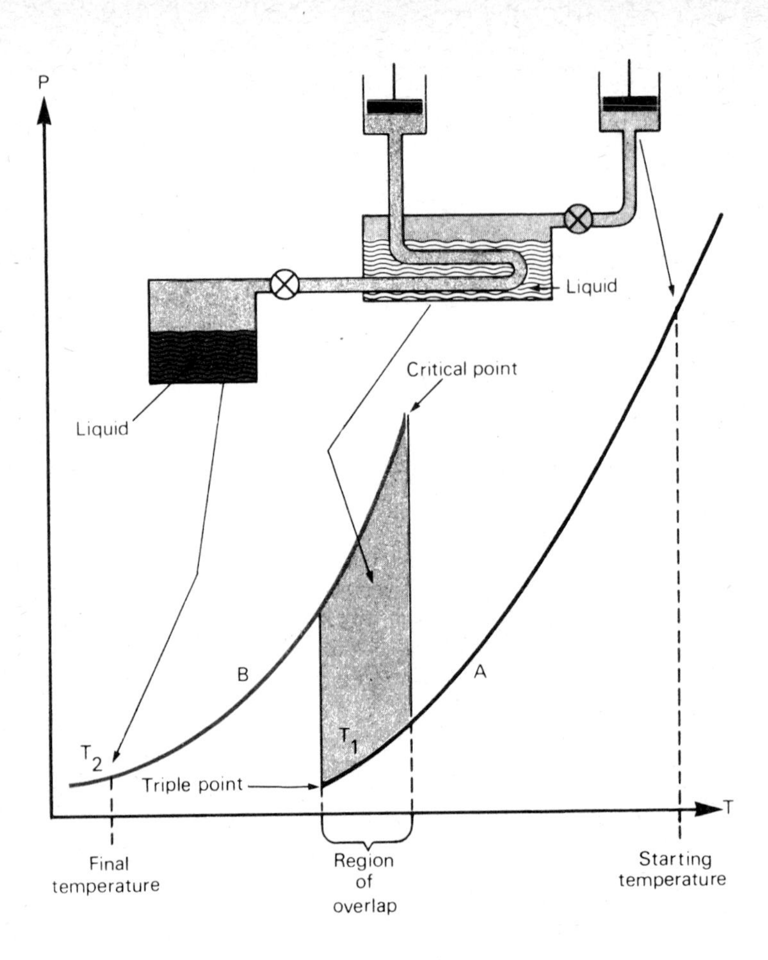

are always the same wherever they are measured.

Returning now to the simple case of a substance enclosed in a container and therefore under its own vapour pressure only, the diagram shows that changes in pressure must result in corresponding changes of temperature. For instance, it now becomes clear why Wroblewski and Olszewski reached a lower temperature than Cailletet. They reduced the vapour pressure of their liquid ethylene bath whereas Cailletet kept it boiling under atmospheric pressure. It also follows from the diagram that any substance whose critical point is above room temperature can be liquefied merely

2·6 A cooling 'cascade' using two substances with different critical data. It is essential that the regions of liquid-gas equilibrium should overlap to some extent.

by the application of pressure. If the pressure is subsequently reduced, the liquid will cool along the (broken) vapour pressure curve. This is the principle according to which most household refrigerators work. If a sufficiently strong pump for reducing the vapour pressure is used, the substance can also be cooled below the triple point so that the solid state can be reached in this manner. For most practical purposes this is not desirable because, once the working substance is frozen, it cannot flow through the refrigeration cycle.

Moreover, one can shunt two or more such refrigeration cycles in series (figure 2·6). Using a substance A which can be liquefied at room temperature by compression, we can evaporate it at some low temperature T_1. Another gas B, whose critical point is well below room temperature but above T_1, can now be liquefied by compression at T_1 and subsequently evaporated at a still lower temperature T_2. This method has been given the self-explanatory name of 'cascade' and it was such a cascade with which Pictet liquefied air in 1877, using sulphur dioxide in his first cycle, carbon dioxide in the second one and oxygen in a third. At least that was his belief at the time of the experiment, which was designed more elaborately and with probably better understanding of the relevant factors than that of Cailletet. However, a later analysis of his experiments makes it unlikely that in the third cycle oxygen was actually liquefied under pressure and it seems that the jet of liquid which he observed issuing from his machine was oxygen liquefied in the same way as in Cailletet's arrangement – by the release of pressure.

Even if Pictet's first cascade only worked in part, he and others showed subsequently that it can be made to function most successfully and serve as an efficient means for liquefying oxygen or air. A cascade employing three successive cycles of methyl chloride, ethylene and oxygen was constructed between 1892 and 1894 by the Dutch physicist Kamerlingh Onnes at his laboratory in Leiden. To this was added a fourth cycle for the condensation of atmospheric air. The whole installation which, at the time, was the most outstanding example of low temperature engineering, produced 14 litres of liquid air per hour and remained in operation for many years, until after Onnes' death in 1926. It formed the basis of the famous Leiden laboratory which today bears Kamerlingh Onnes' name and which was to dominate low-temperature research for more than three decades. This, however, is a later phase of our story.

Immediately after the liquefaction of air, Cailletet attempted in the same manner to liquefy hydrogen, but all his efforts failed. Pictet, too, tried his hand at it and here, unfortunately, his enthusiasm outpaced his scientific caution. He felt so certain that his apparatus was producing liquid hydrogen that he made some statements which later proved to be quite untenable. On the basis of a mistaken chemical prediction, he expected liquid hydrogen to have metallic properties and he reported having seen a steel blue jet of liquid hydrogen which struck the wall of his apparatus with a metallic clatter.

In 1884 Wroblewski and Olszewski, now working separately in Cracow, each set up experiments to liquefy hydrogen by Cailletet's expansion method. Both saw a slight mist which each hoped to be droplets of liquid hydrogen but neither was free of the suspicion that it might have been some impurity in solidified form. However, the days of Cailletet and Pictet were now over and anyone claiming seriously to have liquefied hydrogen was expected to demonstrate something more definite than a transient whiff of mist. The failure of these Cracow experiments marks the end of the brute force and hit-or-miss attempts at gas liquefaction.

The emphasis behind the experiments had now completely changed. The myth of 'permanent gases' had been exploded and Lavoisier's prediction had been proved correct. After Cailletet's liquefaction of oxygen and nitrogen nobody doubted seriously that hydrogen, too, could be turned into the liquid state. It had ceased to be a problem whether this last remaining gas could be liquefied and had instead become the question of the temperature at which this would occur. An important chapter of scientific enquiry had been brought to successful conclusion; the inter-relation of the three states of aggregation had been understood. For any substance a chart could be drawn up – showing solid, liquid and gas – which, except for minor deviations, would always look like the diagram of figure 2·5.

The new chapter which was beginning in the late 'eighties was the quest for low temperatures. Up to then the making of low temperature had been merely an aid to the conquest of the permanent gases and, as we have seen, the part played by these low temperatures was not always well understood. Then, after the first excitement over the liquefied gases had died down, the low temperature aspect took the stage. When the Polish physicists first showed how to maintain liquid oxygen in the laboratory, it began to be realised that science had set out on the path to absolute zero and that in one fell swoop two-thirds of the way had already been covered. The few grams of the bluish liquid inside an apparatus at Cracow were the first outpost in a vast new territory which as yet was completely uncharted. From then on progress took place along two lines which, although depending on each other all the time, had different objects. One was the march towards absolute zero, soon to become an exciting race, and the other the investigation of matter in the newly-opened field of low temperatures. A third purpose was soon to be added: the technological use of the new scientific discoveries.

The first theoretician of low-temperature research was the Dutch scientist Johannes Diederik van der Waals who, in 1872, at the age of thirty-five, gave an interpretation of Andrews' results in terms of molecular physics. Kinetic theory regards a gas obeying Boyle's law as an assembly of molecules which are so small that their size can be neglected in relation to the space which they occupy. The only relevant feature, apart from their mass, is the velocity with which they move through space and with which they collide. Since these collisions are fully elastic, we may expect the physical laws describing the behaviour of such an 'ideal gas' to be simple, as indeed they are. We have seen earlier that heating by

2·7 Van der Waals' modification of Boyle's law takes into account the space occupied by the molecules themselves and the forces between them. It yields a good approximation to the isotherms observed by Andrews (figure 2·4).

compression and cooling by expansion can be explained in an easy and straightforward manner by means of this simple model. When designing a cooling engine which relies on expansion by doing work against a piston or in a turbine, the laws of Boyle and of Gay-Lussac are all that is needed. However, these laws tell us nothing about liquefaction. If they were always obeyed, an expansion engine could be built which, starting its operation at room temperature, would eventually take us all the way to absolute zero. At the time when people believed that some gases were permanent, this feat must have appeared feasible, although we do not know whether anyone ever considered the possibility. Although van der Waals published his famous paper 'On the continuity of the gaseous and liquid state' five years before oxygen was liquefied, he was already convinced that no gas would be permanent. His aim was to find the reasons for the breakdown of Boyle's law first noted by his countryman van Marum and then investigated so thoroughly by Andrews. Since water vapour will, on cooling, condense first into liquid water and then freeze into ice, the concept of gas molecules possessed only of energy of motion requires amplification.

The two aspects in which the simple kinetic model of a gas cannot be adequate are obvious; the molecules must have finite size and they clearly exert forces upon each other. Van der Waals took account of these two omissions by writing down the gas laws with $(P+a/V^2)$ instead of the pressure P, and $(V-b)$ instead of the volume V. His first amendment takes account of the fact that the attractive force between molecules will bring them closer together and will thus have the same effect as an additional pressure governed by the constant a. This 'pressure' must be the stronger, the closer the molecules are together, hence he divides a by V^2. The second amendment reduces the total volume by the constant b, which represents that part of it which is occupied by the molecules themselves.

The new equation, which instead of the old $PV = \text{const.}\ T$ now reads $(P+a/V^2)\ (V-b) = \text{const.}\ T$, has, when plotted, a peculiar

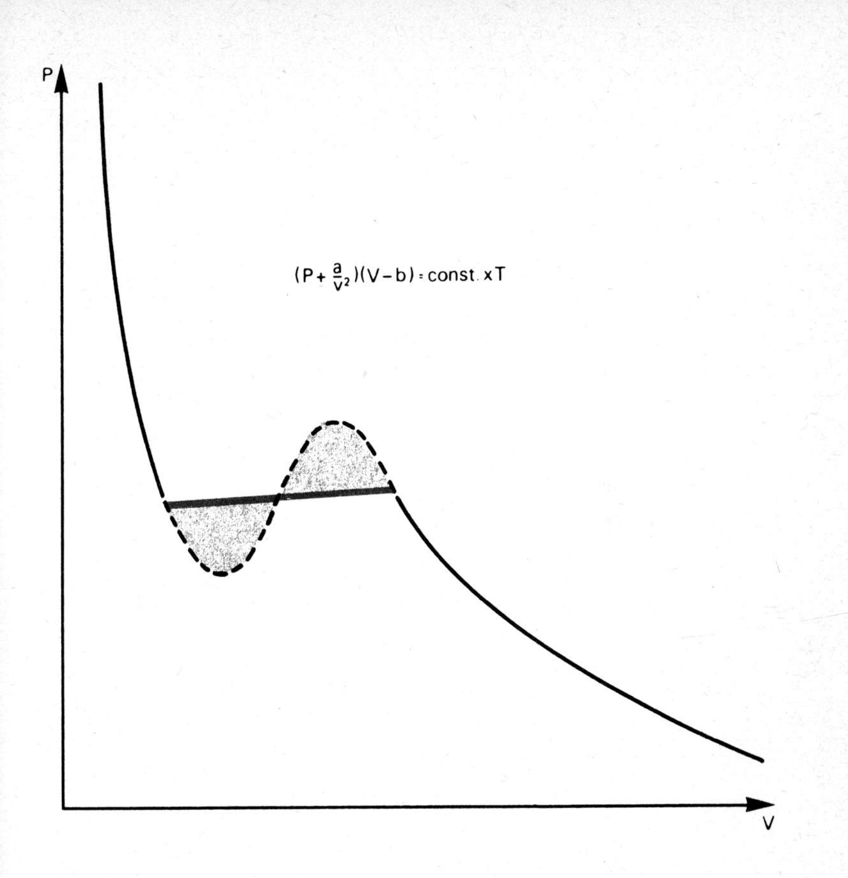

wiggly shape (figure 2·7) which, however, is not unlike the isotherms measured by Andrews. There is no flat portion but it has to be noted that in the wiggly region van der Waals' equation has for any given pressure three solutions for the volume. A straight line, joining any of these three solutions will then result in a curve which very closely resembles Andrews' isotherms. It is, moreover, possible to choose the correct three values by a not too laborious calculation. It soon had to be realised that the nature of the intermolecular forces is far too complicated to be expressed rigorously by such a simple formula as that of van der Waals, and today it has become clear that no simple equation of state can have general validity. However, these difficulties only arise when we are looking for very accurate values of the isotherms and there is no doubt that in its broad concepts van der Waals' approach was correct. Once we are not concerned with finer detail, the van der Waals equation represents remarkably faithfully a complex situation. In addition, the part played by the cohesive forces in the changes from one state of aggregation into another have become clear. These forces can be neglected at high temperature when they are completely overshadowed by the high energy of motion. As the gas is cooled this kinetic energy decreases and the intermolecular forces make themselves felt in deviations from Boyle's law. Finally, at sufficiently low temperatures they will outweigh the decreasing kinetic energy and the molecules instead of bouncing back after each collision will be held together. The liquid state is reached. In it the molecules still have a certain freedom of motion but never get very far away from each other. At still lower temperatures the kinetic energy becomes so small that even this freedom is lost, and from now on each molecule will forever be bound to its neighbour and the whole assembly of molecules is rigidly frozen into the solid state.

The great virtue of the van der Waals equation is that, with the reservations just mentioned, it will apply to any substance. Only the two coefficients a and b differ from case to case. Once these have been determined from measurement of one or several isotherms, the whole diagram of state for the substance can be predicted. In particular, the critical temperature is near enough given by $0{\cdot}15\ a/b$ so that knowledge of the two coefficients for a gas permits the prediction of its critical point. Since the van der Waals equation is also valid in the region above the critical temperature, a and b can be found from careful measurements of isotherms well above the region of possible liquefaction.

After the disappointing result of 1884, Wroblewski devoted most of his remaining four years of life to the determination of the coefficients a and b for hydrogen. After his death the manuscripts of his recorded observations and conclusions was submitted by one of his pupils to the Academy of Sciences in Vienna. The results had not been encouraging. He estimated the critical temperature of hydrogen to be at about 30 K and, as later work was to show, his conclusion was correct. On the other hand, the lowest temperature which can conveniently be reached with liquid air is about 55 K. There was thus no hope of obtaining liquid hydrogen by compressing the gas at liquid air temperatures. A gap of 25 degrees remained between the vapour pressure curves of air and hydrogen with no other gas available which could serve in liquefaction by means of a cascade. Progress had come to a dead stop.

Seven years later, in 1895, Hampson in England and Linde in Germany, again independently and simultaneously, hit upon a new method of gas liquefaction which opened the door to further progress towards absolute zero. It proved immediately successful owing to its simplicity as far as the engineering problems are concerned. The underlying physical principle is, unfortunately, not nearly as simple. The method is based on observations made by Joule and Thomson (later Lord Kelvin) more than forty years earlier. In connection with his work on the transformation of energy, Joule became interested in the expansion of a gas under conditions where no work is done in this expansion. When discussing Cailletet's experiment we have seen that even if no piston is part of the arrangement, work may still be done as, for instance in his case, by lifting up the earth's atmosphere. Looking at the kinetic picture, it is quite clear that work is always done when the molecules of the expanding gas bounce back – either from the piston or from their neighbours – with less speed than they had originally. In Joule's experiment the gas was simply allowed to expand into a larger volume and under these conditions no heat should be evolved because no work is done anywhere in the arrangement. His first experiments were inconclusive but in 1852 he devised, together with Thomson, a more sensitive experiment. They made air flow along a tube in which there was a porous plug. Since the plug forms a resistance to gas flow, the pressure of the air in front of it was higher than behind it. The air thus expanded in passing through the plug but without doing work. In the collisions which take place as the air seeps through the fine channels in the plug, the molecules would nowhere bounce back with decreased speed.

The results were surprising in two respects. First of all, air as well as its constituents, oxygen and nitrogen, cooled slightly in this expansion, as did most other gases investigated by Joule and Thomson. The second surprise was that hydrogen, the only exception to this rule, showed a heating effect instead. Later work then showed that the cooling effect increased with falling temperature and that, at sufficiently low temperatures, even hydrogen cools when expanded in this manner.

Since for an ideal gas, that is, one which obeys Boyle's law, we cannot expect any heating effect at all, the obvious conclusion is that what Joule and Thomson observed must in some way be due to the same features which are responsible for liquefaction. The whole problem could only be solved after the work of Andrews and van der Waals. Andrews' diagram (figure 2·4) shows indeed that well above the critical temperature the isotherms are still far from being rectangular hyperbolae. They show a twisted appearance which is a definite sign for the approach of the liquid state at lower temperatures and is due to the same reasons – the intermolecular forces and the size of the molecules themselves. As the gas is expanding through the porous plug, work is in fact done, though not external work against a piston or against the atmosphere. The work in this case is expended internally by pulling the molecules apart, by overcoming the attraction forces between them.

However, the van der Waals equation does not take account only of the intermolecular forces; it also makes allowance for the size of the molecules and this term of the equation acts in the opposite direction. The finite size of the molecules has the tendency of keeping them apart. Hence there is also a heating effect. We can now understand why the temperature change found by Joule and Thomson is a cooling in some cases and a heating in others. The two corrections contained in the van der Waals equation act in opposite directions and the resulting thermal effect can therefore

2·8 In the Hampson–Linde cycle the expansion engine of figure 1·2 is replaced by a valve where cooling occurs due to the Joule–Thomson effect.

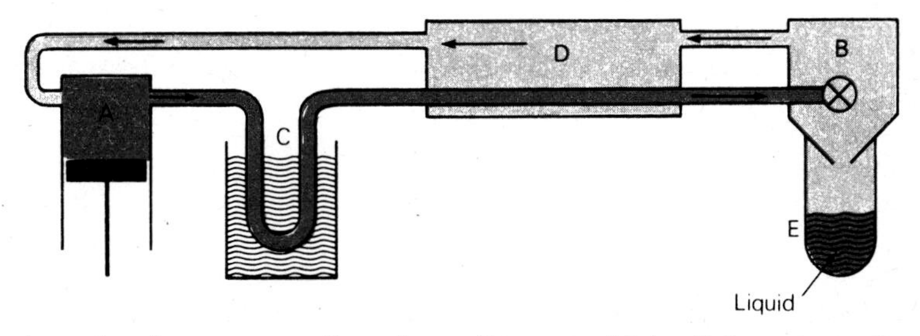

be a heating or a cooling, depending on which of the two is the dominant one. It is fortunate for the making of low temperatures that as liquefaction is approached, the cooling effect is always predominant.

Compared with the cooling which can be obtained in an expansion engine, the Joule–Thomson effect is usually very small. On the other hand, the cooling device proper is simplicity itself – a plug or nozzle through which the gas is expanded – and it does not require moving parts at low temperature. With Siemens' cooling cycle and his heat exchanger, published almost forty years earlier, it was only a question of time before somebody would get the idea of substituting for the expansion engine a Joule–Thomson plug. The patents for such a device were filed independently by Hampson in England on the 23rd May 1895 and by Linde in Germany on the 5th June of the same year. The account of Linde's first experiment a few months later is quite dramatic. Since the heat exchanger was very heavy, it took no less than three days to reach the final temperature and the first appearance of liquid air. In the two intervening nights it warmed up again to some extent owing to poor heat insulation.

The operation of the Hampson–Linde cycle is very simple (figure 2·8). The gas is compressed in *A*, passes through a heat exchanger *D*, and is then expanded through the nozzle *B*. At this point a small lowering in the temperature takes place and the expanded gas thus cools the incoming gas in *D*. As this goes on, the temperature at the nozzle will steadily decrease until at last liquefaction takes place and that fraction which is liquefied collects in E. Comparison between figures 1·2 and 2·8 shows straight away how similar the two cycles are, the only difference being the substitution of the nozzle for the expansion engine. This substitution means, however, that a large loss of cooling efficiency and higher pressures as well as greater quantities of gas passing through the cycle were the price which had to be paid for technical simplicity.

From 1895 on the industrial application of low-temperature physics has never looked back. The foremost aim became the extraction from the atmospheric air of oxygen which is needed in large quantities in steel making and other technological processes. The separation of nitrogen from air and of hydrogen from water gas also are of great importance. The Hampson and Linde patents ultimately led to the great gas-separation industries in England, Germany and the United States, and they were supplemented at the turn of the century by the patent of Claude in France who by then had turned the expansion engine into a commercial proposition.

3 London 1898

On the 1st of September 1894 *The Times*, London, published an article on the approach to absolute zero. It was based on an interview with Professor (later Sir James) Dewar at the Royal Institution in London and mainly dealt with what then appeared to be the last possible step towards the ultimate goal: the liquefaction of hydrogen. Dewar, who was then fifty-two, had by his brilliant experiments moved into the centre of the low temperature stage. Four years later he was to achieve his purpose, though in a very different manner from that foreshadowed in the interview. At that time his hope of bridging the gap between the temperatures of liquid air and the critical point of hydrogen was based on 'the construction of a new substance' which would have a convenient critical point at, say, 80 K. Hydrogen, with a ten per cent admixture of nitrogen, was expected to serve the purpose and Dewar experimented with it. Today we know that the solubility of nitrogen in liquid hydrogen is far too small to be effective but Dewar evidently was not sure whether perhaps he had liquefied hydrogen after all. In any case he was cautious and so *The Times* said: 'Professor Dewar will not declare that he has had pure liquid hydrogen in one of his vacuum vessels, although what this liquid can be except hydrogen it is impossible to say.'

James Dewar was a short man with a versatile mind, uncanny experimental skill, and possessed of an artistic but irascible temperament. The youngest of seven sons of a Scottish innkeeper, he fell through the ice at the age of ten and for several years afterwards was in poor health. It was then that he developed his great manual dexterity by spending much time with the village joiner who taught him to make fiddles. One of these which was played at his golden wedding bore the inscription 'James Dewar, 1854'. Later he went to Edinburgh University where, after his studies, he held a lectureship in chemistry, and at the age of thirty-three he was appointed Jacksonian Professor of Experimental Philosophy at Cambridge. From all that is known it must appear that Dewar and Cambridge did not take kindly to each other and when, two years later, he was offered the Fullerian Professorship of Chemistry at the Royal

Institution in London, he accepted with pleasure and stayed there until his death at the age of eighty-one, working to the last.

Dewar's interest, as is shown by his scientific publications, ranged over a very wide choice of subjects but the rest of his researches are completely overshadowed by his outstanding contributions to the field of low temperatures. At the Royal Institution he felt free from academic duties and university regulations, and the artist in him was ever aware of the ghosts of Davy and Faraday. Hence, when one year after his taking office the liquefaction of oxygen was announced, he felt that the spirit of Faraday commanded him to follow up the earlier pioneer work on gas liquefaction at the Institution. He lost no time in obtaining from Paris a Cailletet apparatus and within a few months, in the summer of 1878, he demonstrated the droplets of liquid oxygen to his audience at one of the famed Friday Evening Discourses. Before demonstrating the apparatus, 'handsomely presented to the Royal Institution by Dr. Warren de la Rue', Dewar gave a history of gas liquefaction, starting significantly with Faraday's letter to Dr. Paris on the liquefaction of chlorine gas.

This discourse was the first of a brilliant series of demonstrations, spread over more than three decades and culminating in the impressive experiments with liquid hydrogen. The rare combination of experimental skill and artistic flair, backed by deep scientific understanding, made Dewar an incomparable showman. A large canvas at the Royal Institution shows him – in evening dress – demonstrating the properties of liquid hydrogen to a large gathering. His lectures were all the more impressive because he usually demonstrated in them his own recent researches. They were social occasions at which the public could witness the drama of current scientific progress and the lecture room became a theatre in the truest sense. As on any stage, the audience only saw the polished performance, hardly realising the infinite care which had gone into planning and rehearsals.

To repeat Cailletet's experiment with a ready-made apparatus is one thing; to set up low-temperature facilities suitable for research is a much more difficult proposition. It took Dewar another six years to arrive at the second stage, and even then he had not gone much further than, in the preceding year, had Wroblewski and Olszewski, to whose 'splendid success' he referred at the beginning of his discourse. Dewar's chief improvement was a modification which permitted the experiments to be seen by a large audience. However, researches on the chemical behaviour of liquid oxygen and the announcement by the President of the Royal Society, on the 27th May 1886, that he had witnessed Dewar solidifying oxygen, indicated that his facilities must have increased. Indeed, in the same year Dewar described his apparatus, the design of which was a distinct advance on the Cracow equipment. Strangely enough the description is appended to a paper which deals with researches on meteorites, a fact which was to lead to some repercussions a decade later.

The function of the Royal Institution and Dewar's love of demonstration experiments made it imperative that the liquefied gases should really 'boil quietly' in a test tube as demanded by Jamin. To achieve this, two things are necessary. First, a sufficient quantity of the liquefied gas must be produced, and, secondly, steps must be taken not to evaporate it again immediately. The first problem was solved by the Polish scientists and they also went some way in dealing with the second. Cailletet already had found that observations in his first experiment were hampered by the frost which formed on the outside of his glass tube. In his later work he therefore surrounded this tube by a second one, fitted on to the inner tube with stoppers and filled at the bottom with calcium chloride, a drying agent. This arrangement had the effect of enclosing the experimental tube with a space free of water vapour so that no condensation could take place. In the early Cracow experiments this device was copied and for the first time liquid oxygen could be seen boiling in a test tube, albeit for a short time because the influx of heat soon evaporated it again. In the following year, 1884, an improvement was made by drawing off the cold ethylene vapour over the outer wall of the glass vessel containing the liquid

3·1 Stages in the development of the vacuum flask.

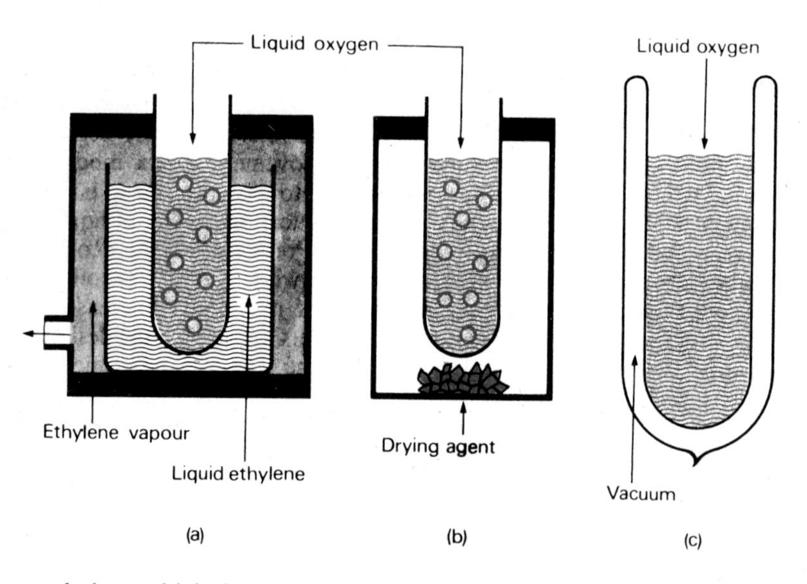

ethylene which, in turn, shielded the liquid oxygen from inflowing heat (figure 3·1a). This was the system adopted at the same time by Olszewski and Dewar.

It was at this time that a new word was added to the scientific vocabulary. The specially designed container in which the liquefied gas was kept for observation and research became known as a 'cryostat' (from the Greek cryos = cold) and the art of producing cold was called 'cryogenics'. Soon the cryostats ceased to be part of the liquefaction apparatus. Instead, liquid was drawn off from the expansion vessel through a tap and allowed to run into the cryostat proper, which could then be detached. This made handling and experimentation much easier. The cryostats of this period consisted of a test tube which contained the liquid oxygen and an outer glass beaker into which the test tube was fitted with a stopper (figure 3·1b). At the bottom of the beaker a drying agent was placed which absorbed the water vapour in the space between the glass walls and prevented frosting. In his discourses Dewar could thus carry into the lecture theatre liquid oxygen which had been prepared shortly beforehand and demonstrate its properties to his audience. In the ten years since the first liquefaction, experimentation with liquid air and liquid oxygen had made enormous strides but there was one rather disquieting feature which indicated a further difficulty in the approach of absolute zero.

In order to turn liquid into vapour a quantity of heat is required and it is the existence of this 'latent heat' of vaporisation which makes it possible to keep a liquefied gas at all. Each cubic centimetre of liquid oxygen requires a heat influx of 300 joules into the cryostat for its evaporation and the less heat is allowed to flow into the cryostat, the longer the oxygen will therefore keep in the liquid state. The trouble is that the latent heat of oxygen is about four times smaller than that of water which, as was explained by Trouton, is due to the fact that – on the absolute scale – oxygen boils at a temperature four times lower than water. Applying Trouton's rule, it had therefore to be expected that the latent heat of hydrogen would be still lower, and possibily more than four times less than that of oxygen. This rough calculation showed only too clearly that, even if it were possible to liquefy hydrogen, the type of cryostat in existence would be quite incapable of holding it for any length of time.

It seems that Dewar solved this important problem sometime late in 1892, because in a discourse on the 20th January 1893 he demonstrated to his audience his famous vacuum vessel in a form so perfect that it has remained unchanged to this day. As we have just seen, the double-walled cryostat had been in use for a number of years before, but only water vapour had been excluded from the interspace. Dewar transformed it by excluding the air as well (figure 3·1c) and he quotes as inspiration for this ingenious step one of his own publications twenty years earlier in which he had used the vacuum for heat insulation in a calorimeter. In his lecture he demonstrated dramatically the superiority of his invention over the old type of cryostat by first showing liquid oxygen quiescent like water in a vacuum vessel and then breaking off the tip where the vessel had been sealed off. As air entered the space between the walls, the liquid oxygen began to boil away violently. At the same lecture he also exhibited vacuum vessels with silvered walls in which, in addition, the heat loss by radiation was greatly reduced.

There was more to the invention of the vacuum vessel than just the brilliant idea. The making of these vessels required very skilled glass-blowing and careful heat treatment of the finished flask so that it would not shatter when the interior was suddenly cooled to the temperature of liquid air. Some of the vessels shown at the discourse were of rather complicated design and one cannot help feeling that this was perfection preceded by many trials and failures. Of these difficulties Dewar, as usual, said nothing. He dealt, however, at length with the ingenious methods which had enabled him to attain the high vacua required for his flasks.

The Dewar vessel was an enormous step forward in the management of liquid air temperatures and in the attainment of still lower ones. In particular, the vastly improved storing property of the vacuum flask now permitted experimentation with pints instead of with cubic centimetres of liquefied gases. At the same discourse Dewar dwelt on the financial difficulties of research, and there was a note of relief when he said: 'While suffering great anxiety on the question of expenditure, the Goldsmiths' Company came forward with a handsome contribution of £1,000 to continue the work with improved apparatus.' Three days later, when reporting on the discourse, *The Times* returned to this theme, referring prophetically to a new law discovered by scientists: the law of increasing expense.

As frequently happens in the case of outstanding inventions, other claimants are put forward, and for the vacuum vessel the Germans invoked the name of Weinhold, who is said to have used vacuum insulation before Dewar, and the French called it *vase d'Arsonval* because the latter had employed vacuum insulation for biological purposes in 1888. There can, however, be no doubt that if either of these scientists had realised the importance of vacuum insulation for preserving liquid oxygen, they would have made use of it. It was Dewar's genius in applying it that set a new pattern for cryogenic experimentation. His use of vacuum insulation in calorimetry twenty years earlier in any case preceded the work of both Weinhold and d'Arsonval by a comfortable margin.

This was not, however, the only question of priority in which Dewar found himself involved, and a few years later there occurred another of those controversies which punctuated his scientific career and to which he appears to have been unduly prone.

In the February issue of the *Philosophical Magazine* of 1895 there appeared an article by Olszewski which in twenty-five pages gave a potted version of all his low-temperature work. In the introduction Olszewski gave two reasons for this publication, the first being that his work was scattered over a number of Continental journals and therefore might not be easily accessible to English readers, and the second that Dewar had repeated his work without even quoting him. Whereas the first argument could pass as legitimate, if unusual, the second was both unjustified and ill-advised. Unjustified, because Dewar had, in fact, anticipated some of Olszewski's work, and ill-advised because Olszewski had sadly underestimated Dewar's fury and venom. Olszewski's main complaint was that Dewar had not taken note of his own apparatus for liquefying large amounts of oxygen in 1890 and it seems that he had never seen Dewar's paper of 1886 on the meteorites in which the gas liquefaction installation at the Royal Institution had been described. It was therefore quite silly to say that Dewar had done nothing more than repeat the work at Cracow. Dewar put this right in his reply and then he hit back where it hurt most. He evidently knew a thing or two about Olszewski's falling out with Wroblewski right at the start of the Cracow experiments, and so he wrote: 'Personally I am delighted to see an English edition of Professor Olszewski's papers. There is, however, one serious omission which I trust the Editors of the *Philosophical Magazine* may remove before long. We want in this country a reprint of the splendid papers of the late Professor Wroblewski. Until this is done it will be impossible for the scientific public to decide on many of Professor Olszewski's claims for priority.' A year later, in a long paper on the continuation of his low temperature work, Dewar returned to the theme with gusto and fresh vigour. He advises his readers to read a paper published by Wroblewski in 1885 'and make themselves generally acquainted with the work of this most remarkable man, before coming to hasty conclusions on claims of priority brought forward by his sometime colleague'.

The apparatus and experiments described in the rest of this paper are of considerable importance; they laid the foundation of Dewar's greatest triumph and bore the seeds of two more acrimonious controversies. The first illustration shows the apparatus used in the Royal Institution for air liquefaction which clearly makes use of the Joule–Thomson expansion through a nozzle as cooling device and uses a heat exchanger in conjunction with it. It has to be noted that this paper was published early in 1896, only a few months after the patents of Hampson and Linde. Linde is mentioned in the text but Hampson's name only appears in an angry footnote against a professor of the Royal College of Science who ascribed the vacuum vessel to Cailletet. The professor's paper was entitled 'L'Appareil du Dr. Hampson pour la Liquefaction de l'air'. On the whole Dewar was at pains to point out that his own apparatus was useful but did not really involve any new principle since the heat exchanger had already been described by Siemens as early as 1857, and he also says that 'the experiments of Joule and Thomson . . . are well known'.

Among the experiments described in this paper the most important ones are those in which hydrogen gas is expanded in a similar type of arrangement to that used for air liquefaction, again making use of a Joule–Thomson nozzle and a heat exchanger. Dewar noted, as was to be expected, that no cooling took place when the hydrogen was fed into the apparatus at room temperature. However, when precooling the gas with liquid air he observed a cooling effect, though there was no sign of liquefaction. He demonstrated the very low temperature of the jet of hydrogen gas issuing from the nozzle by playing it on liquid oxygen; the oxygen froze into a hard solid of pale blue colour. Dewar estimated the temperature of the jet as between 20 and 30 degrees above the absolute zero and he points out that the properties of matter could now be studied at these very low temperatures. That the liquefaction of hydrogen itself could be achieved, he was now certain. When discussing the prospects he says: 'These difficulties will be overcome by the use of a differently-shaped vacuum vessel, and by better insulation. That liquid hydrogen can be collected and manipulated in vacuum vessels of proper construction cannot be doubted.'

In the following two years he published a large amount of research on the properties of matter at liquid air temperatures but it is clear that behind the scenes a full-scale attack on the liquefaction of hydrogen was under way. After all, Olszewski in Cracow was working hard on the same problem and a new and powerful competitor had arisen in Holland. Thanks to his large cascade-type air liquefaction, Kamerlingh Onnes in Leiden had cryogenic facilities which were probably superior to those at the Royal Institution and Dewar knew that Onnes, too, was preparing for hydrogen liquefaction. Then, on the 10th May 1898 Dewar reached his goal: he had produced 20 cubic centimetres of liquid hydrogen boiling quietly in a vacuum vessel. The announcement was made at the Meeting of the Royal Society on the 12th May.

Neither then nor at any future occasion did Dewar give a description of his liquefier. He simply mentions that at the first successful attempt the hydrogen gas was precooled to −205°C (68 K) and that it was passed through his apparatus at the rate of about 12 cubic feet per minute. After an unspecified time liquid hydrogen began to collect and the 20 cubic centimetres were obtained in about five minutes running time, after which the apparatus became blocked by impurities in the hydrogen gas. Dewar observed the liquid as colourless with a clearly defined meniscus and a rough measurement showed that the density is quite small; liquid hydrogen has only one-fourteenth the weight of water. Dewar was, of course, chiefly interested in the temperature which he had reached

Cailletet

Pictet

Wroblewski

Olszewski

Andrews

Dewar

Kamerlingh Onnes

Nernst

Planck

Einstein

Giauque

Debye

but here he encountered a peculiar difficulty: the electrical thermometer installed in his apparatus gave a ridiculous reading and had evidently ceased to function. New laws of nature, of which Dewar had as yet no inkling, were heralding the approach to absolute zero.

While at this first liquefaction Dewar was unable to determine the boiling point of liquid hydrogen, he could show that the temperature reached must be very low. Two sealed glass tubes, one filled with oxygen gas and the other with air, were lowered into the cryostat containing liquid hydrogen and both substances became immediately solid. The electrical thermometer used in his first experiment was a thermo-couple, a device which records a voltage between two junctions of different metals, one of which is held at the temperature to be measured. To his surprise no appreciable voltage was recorded to account for the drop in temperature. Subsequently another type of electrical thermometer was used which relies on the change of resistance in a platinum wire with varying temperature. In his earlier experiments with liquid air and oxygen Dewar had found that this resistance dropped at the same rate as the temperature was lowered. There was also such a drop in resistance when the thermometer had been cooled to the boiling point of hydrogen but the temperature read from it was suspiciously high; 35 K. Here again something strange was happening; the drop was too small, indicating that the law of resistivity change, too, deviated from the normal pattern as absolute zero was approached. Finally, a gas thermometer was tried, an instrument which relies on the law of Gay–Lussac which we have discussed earlier (p. 11). Dewar used hydrogen under low pressure in his gas thermometer in order to come as near as possible to the conditions for a perfect gas but some correction had to be applied. He deduced pretty accurately the correct boiling point as about 20 degrees above absolute zero.

Hardly had Dewar announced the successful liquefaction of hydrogen when, two weeks later, a Letter to the Editor appeared in *Nature* which was signed by Hampson who complained that he had not been given any credit for his part in the experiment. Hampson claimed that before his patent was applied for in May 1895 he had gone to the Royal Institution late in 1894 to see Dewar's assistant, Lennox. Now, according to Hampson's story, he had told Lennox about his invention, suggesting that Dewar had achieved his successes, both in the liquefaction of air and of hydrogen, on the basis of information given by Hampson to Lennox at this interview. This letter brought an angry and stinging reply from Dewar, saying that he would have succeeded in the same way and at the same time if Hampson had never existed. After this Hampson retorted, going again over much the same ground, and so did Dewar. The editors of *Nature* seem to have derived some enjoyment from this undignified quarrel because by the 4th August they had published no less than four letters from Hampson and four replies from Dewar. Then the whole thing died down and it is now impossible to say whether the contestants got tired or either the editors or the holiday season put a stop to it. There is something odd in this controversy and it is impossible to say now what exactly was fishy about it. What did Hampson want at the Royal Institution, and why did he talk to Lennox instead of seeing Dewar? Why, on the other hand, did Dewar mention Linde in his paper of 1896 on air liquefaction and not Hampson, although his footnote shows that he knew of the latter's work? It is quite possible that Dewar was thinking on the same lines as Linde and Hampson did, independent of each other, and in this case Hampson's telling Lennox about his ideas would have been an embarrassment to Dewar. It has to be remembered that Dewar had succeeded brilliantly in liquefying air with Joule–Thomson cooling well before Hampson's machine ever worked.

Unedifying as the row was, it does not seem to have damped Dewar's spirits. In the year after the first liquefaction of hydrogen he achieved his next, and last, triumph on the road to absolute zero: the solidification of hydrogen. His first attempts at obtaining the solid phase by pumping off the vapour over the liquid had failed and he immediately realised that the stray heat influx into his cryostat was too large to be counterbalanced by the cold which he could obtain in evaporating the liquid. The worst fears that conditions for further cooling would become progressively more unfavourable as absolute zero was approached were not only fulfilled but even surpassed. As mentioned earlier, it had to be expected from Trouton's rule that the heat required to evaporate liquid hydrogen might be only a quarter or a fifth of that needed for liquid oxygen. This would make it four to five times as difficult to maintain liquid hydrogen, demanding much better thermal protection of the cryostat. It turned out that this heat of evaporation of hydrogen is still much smaller than even Trouton's rule indicates. The reason for this added and unexpected difficulty lies in the same deviation from the ordinary laws of physics which had falsified the readings of Dewar's electrical thermometers.

Dewar eventually overcame the difficulty of the low heat of evaporation by placing his vacuum vessel containing the liquid hydrogen to be solidified inside another vacuum vessel filled with liquid air. The outer vessel then acted as a cold shield which minimised heat flow into the central part of the cryostat. As the pressure of the pumped-off vapour fell to 5 cm mercury column, a froth appeared in the liquid which in due course began to form a clear transparent mass. Hydrogen had been solidified. As was to be expected from the clear nature of the liquid, solid hydrogen did not turn out to be a metal, although some chemists had predicted this confidently.

Again there was the difficulty of determining the temperature of the triple point. Dewar computed this from his gas thermometer readings as about 16 K, but it later turned out that he had been unnecessarily cautious in his estimate. He had, in fact, approached absolute zero to within 14 degrees. Some further cooling can be achieved by pumping off the vapour over the solid, although no large decrease of the temperature can be expected from this because the vapour pressure soon becomes very small. Here again Dewar underestimated his achievement by believing that 13 K would be the lowest temperature attained with solid hydrogen. How far he actually got is now difficult to say since all depends on details of the experimental arrangement, but he probably reached 12 K or an even slightly lower temperature.

When Dewar carried out his first successful hydrogen liquefaction, he believed that he was taking the ultimate step in the approach of absolute zero and it was due to his own labours that this proved wrong. Hydrogen turned out not to be the gas with the lowest boiling point. The next stage on the road to absolute zero was called helium.

His first communication on hydrogen liquefaction was, in fact, entitled 'Liquefaction of hydrogen and helium' but it soon became clear that what he had taken to be condensed helium were traces of impurity. Helium was quite a new name in the story of gas liquefaction and this is not surprising since it turned out to be so rare a substance that it had for long escaped discovery. Helium is not only rare but it is chemically quite inactive, forming no compounds with other substances, which is the usual method through which a rare element is found. As a matter of fact its presence on earth is so little noticed that it was first discovered on the sun.

In the middle of the nineteenth century Kirchhoff and Bunsen made an enormous advance in the study of matter by analysing the the light emitted from incandescent vapours. They used an instrument, the spectroscope, which measures the individual wavelengths in any source of light. At the total eclipse of the 18th August 1869, the spectroscope was for the first time turned towards the solar corona, a vast envelope of hot gases which only becomes visible when the disc of the sun is obscured by the moon. The total eclipse passed over India and Malacca and all the scientists who observed it noted a bright yellow spectrum line which most of them at first ascribed to hydrogen, or possibly sodium. One of them, Janssen, wondered whether under ordinary conditions when the corona is invisible the spectroscope might nevertheless reveal the yellow line. He tried the next day and was successful. He reported his observation to the French Academy, where his letter arrived on the 24th October, the same day as one from Lockyer in England, stating independently the same result. Both Janssen and Lockyer from the very first had doubts that the strong yellow line which they had seen in the sun was identical with the well-known sodium line, but it was more difficult to be sure about hydrogen. In the end, comparison with observations in the laboratory, mainly by Frankland and Lockyer, left little doubt that the line belonged to an unknown chemical element, and in his presidential address to the British Association in 1871 Lord Kelvin summarised their work by saying: 'It seems to indicate a new substance which they propose to call Helium.'

For nearly a quarter of a century helium remained a gas only observed on the sun. Then, in 1895, Sir William Ramsay was following up his discovery of the new inert gas argon by investigating the gases expelled from pitchblende when this mineral is heated. He submitted them to spectrum analysis and found to his great surprise the bright yellow line seen by Janssen and Lockyer in the sun's atmosphere. Helium had been found on earth, and now the search was on for other terrestrial sources of the new substance. Mineral springs, such as Wildbad and Bath, yielded small amounts of helium and so did gas wells. Helium was also found to be present in the earth's atmosphere but only at a concentration of less than 1 in 100,000.

Dewar soon found out that his claim of having liquefied helium at the same time as hydrogen was erroneous and that the new substance remained, in fact, still gaseous at the lowest temperatures which he could obtain with solid hydrogen. At the height of his triumph he found himself in a strange situation. It was, after all, *not* the last remaining step on the road to absolute zero which he had taken with the liquefaction of hydrogen. His achievement had disclosed another possible stage and from all indications before him he had to conclude that the liquefaction of helium would be an even more difficult proposition than that of hydrogen. Moreover, his competitors, Olszewski in Cracow and Onnes in Leiden were working on the same problem. The race was on but, unless helium had a fairly high critical point, it promised to be a long and difficult road. A critical point above 10 K might be reached by expansion in a Cailletet apparatus and this could be undertaken with a limited amount of helium. On the other hand, for temperatures below this a Joule–Thomson cooling cycle had to be invoked, requiring quantities of helium gas far exceeding those which were available at the time.

Dewar did not enter the race with the advantage which was so near at hand. The world's greatest authority on the rare gases, Ramsay, had his laboratory in easy walking distance from the Royal Institution but, unfortunately, Dewar was not on speaking terms with Ramsay.

It had all started at a meeting of the Royal Society in December 1895 where Dewar was reading a paper on his low-temperature work, full of confidence that he was on the threshold of hydrogen liquefaction. After he had finished, Ramsay got up to say he had just learned from Cracow that Olszewski had successfully liquefied hydrogen and had obtained not only a mist but a fair amount of liquid. Ramsay's statement must have formed a sad epilogue to Dewar's paper. Dewar waited in vain for Olszewski's official announcement until, in May 1898, he himself was able to report the first hydrogen liquefaction to the Royal Society. To Dewar's consternation Ramsay again after his talk repeated Olszewski's claim to priority. Dewar now challenged Ramsay to produce some proof of his assertion and at the next meeting Ramsay had to admit that, in a letter just received by him, Olszewski denied ever having obtained liquid hydrogen in static form. There can be no doubt that, without any justification, Ramsay had grossly provoked Dewar. Dewar was, understandably, annoyed and, since Ramsay failed to publish his retraction in print, Dewar rectified this by publishing the controversy and his own vindication in the Proceedings of the Royal Institution. While he was certainly entitled to do this, it did not improve his relations with Ramsay. Dewar clearly did not care. On the contrary, Ramsay was at a greater disadvantage than Dewar because he needed liquid hydrogen to separate helium from neon and, as things now stood, he could hardly ask Dewar for

3·2 *Overleaf*: Sir James Dewar demonstrates the properties of liquid hydrogen at the Royal Institution. *Key*: 1 and 2 Mr. and Mrs. Alexander Siemens; 3 Lady Dewar; 4 Sir William Crookes; 5 Rt. Hon. Lord Rayleigh; 6 Dr. Ludwig Mond; 7 Sir Oliver Lodge; 8 Sir James Dewar; 9 Mr. J. W. Heath; 10 Mr. R. N. Lennox; 11 Dr. Rudolf Messel; 12 Prof. Sir Francis Galton; 13 Robert Mond Esq.; 14 Commendatore G. Marconi.

facilities at the Royal Institution. However, he was fortunate in having at University College a most able young man, Morris Travers, who, in only two years built a hydrogen liquefier, made it work and published its construction in detail. Travers was at pains to point out that the only reason for building the liquefier was to make experiments with rare gases but at the same time he ostentatiously thanked Hampson for his help and even more pointedly remarked on the simplicity of construction and the small outlay of money required; in his case £35. That was a dig at Dewar who had always been outspoken about the costliness of low-temperature research.

It was clear that Travers and Ramsay, now having liquid hydrogen at their disposal would also have a shot at liquefying helium, and so by 1901 Dewar faced not two but three competitors. Like himself, they all tried the shortcut first. In fact, Olszewski had already tried to liquefy helium by Cailletet expansion in 1896, that is, before the liquefaction of hydrogen. He started at liquid air temperatures and at a pressure of 140 atmospheres. He was probably optimistic when estimating the temperature reached on expansion as 9 K, but, in any case, there was no sign of liquefaction. Dewar, in 1901, started at the triple point of hydrogen (14 K) and 80 atmospheres, and was equally unsuccessful. He, too, estimated his final temperature as 9 K, which was probably pessimistic, and he may have actually descended 2 or 3 degrees below his estimate. Two years later Travers was ready for his attempt, starting from strongly pumped solid hydrogen at a temperature of probably between 11 and 12 K and 60 atmospheres, but he also met with failure. In 1905 it was again Olszewski's turn and he was followed by Kamerlingh Onnes.

By then it had become abundantly clear that the simple Cailletet apparatus with its relatively large heat capacity and the small quantity of helium involved was not leading to helium liquefaction and that there was no escape from the arduous task of setting up a Joule–Thomson liquefier. However, even that might fail if even at the lowest temperatures obtainable with hydrogen helium would still show a heating when passed through a porous plug or expansion valve. Thus, before committing oneself to a laborious and costly project of a full-scale cooling cycle with helium, it was essential to obtain better estimates of its critical point. The first attempt was made by Dewar in 1904 with an ingenious experiment in which he studied the adsorption of helium on charcoal at hydrogen temperatures. This yielded an estimate of 6 K, which was promising. However, one year later Olszewski concluded that the critical point of helium would be as low as one degree above absolute zero, and this seemed hopeless. This depressing result was confirmed in the following year by Kamerlingh Onnes on the basis of a study of helium–hydrogen mixtures. Fortunately, he did not rely entirely on this conclusion, feeling that a reliable estimate could only be obtained from a proper measurement of the isotherms. This was under way in his laboratory and when the results became available after another year, in 1907, they completely changed the outlook. They showed that the critical point must lie between 5 and 6 K; Dewar's estimate had been correct.

Meanwhile Dewar was working hard on the liquefaction cycle. Lennox wanted to build the liquefier entirely in metal but Dewar overruled this, thinking that he must be able to see what was going on. This was probably a mistake, but the greatest difficulty was to obtain a sufficient quantity of pure helium gas. In retrospect it seems likely that a combination of Dewar's skill in low-temperature work and Ramsay's experience in the handling of rare gases would have made all the difference to the success of the project. Instead, Dewar and Lennox built an enormous contraption for separating helium and neon which never worked. Consequently all their liquefaction attempts were bogged down by impure helium, the tubes and valve of their liquefier becoming blocked with frozen neon. Then another disaster occurred. A young workshop technician turned the wrong tap and all the precious store of helium was lost overnight.

In 1908, in a paper entitled 'The Nadir of Temperature' Dewar discusses his difficulties and tribulations in the attempt at liquefying helium but is hopeful in the prospect of success; if only 100 or 200 litres of pure helium were available. Then there is a sad footnote, added in proof to the paper at this point. It reads: 'Helium was liquefied by Professor Dr. Kamerlingh Onnes, of Leiden University, on July 9, 1908.' The race was over and Dewar had lost.

Dewar never got over his defeat. Immediately after the blow he quarrelled with Lennox, and Lennox left. He now rapidly lost interest in low-temperature studies. They still went on to some extent at the Royal Institution but his real interest turned to other problems and particularly to studies of the thin liquid films provided by soap bubbles which continued to intrigue him to the end. He became even more autocratic and quarrelsome than before and eventually fell out with his one remaining scientific friend, Sir

William Crookes. In the end it was only the companion of his childless marriage, Lady Rose Dewar, who stood at his side. Her worship of her husband never faltered, and he was devoted to her. It was she who, after his death, saw to it that his collected papers were published in book form. In spite of his many faults, he was often unexpectedly generous, a lover of music who would not only sponsor young artists but secretly buy up and distribute a large share of the tickets so that they would have an audience.

With Dewar, low-temperature work in Britain came effectively to an end. He left no School. Having preferred the Royal Institution to Cambridge, he turned the Royal Institution into a one-man show, although admittedly a brilliant one. His scientific assistants, Lennox, Heath and Green, were occasionally given a short note of thanks at the end of a paper but their names never appeared in the title. Lennox and Heath each lost an eye working for Dewar, and his successes must have owed much to their devotion, either to him or to the work which he inspired. Dewar's rule in his laboratory was as absolute as that of a Pharaoh and he showed deference to no one except the ghost of Faraday whom he met occasionally at night in the gallery behind the lecture room.

4 Leiden 1908

After Paris, Cracow and London, the scene is shifted to Leiden. In the Low Countries, with their centuries-old tradition of scientific research, the first pioneering work in the field of gas liquefaction had been carried out in the eighteenth century by Martinus van Marum, the Director of the Teyler Stichting at Haarlem. We have already mentioned his work on ammonia as well as that of van der Waals, who provided the theoretical interpretation of Andrews' experiments on the critical point. It was the work of van der Waals which directly inspired the foundation of the most important institution in the history of low-temperature research, the Leiden Cryogenic Laboratory.

In 1882 the University of Leiden appointed to the chair of physics a young man of only twenty-nine whose early work had already given great promise. He came from an old family in Groningen in the north of Holland, and his name was Heike Kamerlingh Onnes. Two years earlier Onnes had been much impressed by one of van der Waals' papers which dealt with the concept of corresponding states. In it the essentially similar behaviour of all substances with regard to the three states of aggregation was postulated. Onnes was particularly intrigued by the prediction of the critical points of as yet unliquefied gases which could be made on the basis of the van der Waals equation. The key to the problem was the measurement of the isotherms (see figure 2·4) from which the constants *a* and *b* could be obtained. However, in order to be useful, these determinations had to be made with great accuracy. It is significant that Onnes selected for his inaugural address as professor 'The Importance of Quantitative Investigations in Physical Science', in which he demanded that the motto *door meten tot weten* (through measurement to knowledge) should be written over the door of every physics laboratory. He maintained that physical observations which at that time were frequently of a descriptive and qualitative nature should be made with the same care and precision as those of astronomy.

Although he was deeply interested in the great revolution of physical concepts and theories which was to take place during his lifetime, his main preoccupation was with the measurements from which these new ideas were to arise. He was essentially an experimentalist with a keen sense for the engineering problems involved in the perfection and proper use of scientific instruments. Without ever becoming a perfectionist, he was acutely aware of the importance of careful planning and organisation for the success of an experiment and he used these ideas on a scale which had never previously been attempted in a physical laboratory. Quite apart from the decisive role which his laboratory played in the development of low-temperature research, it also served as the model for the research institutions of the twentieth century. Kamerlingh Onnes was not only a master of organisational genius, he was also a good and patient diplomatist and a wise man. His great strength was to plan not for tomorrow but for the day after. The secret of the steady output of brilliant work which issued from his institute lay in the fact that each experiment had been thoroughly thought out and prepared long before work on it was begun.

Onnes was probably the first scientist to realise that the complexity of modern research techniques would require a reliable supply of skilled and specially trained assistants. He sensed that the time of the amateur professor was over, who could go into the laboratory and discover the secrets of nature in odd afternoons with the aid of string and sealing wax. In 1901 he founded at the laboratory a school of instrument makers and glassblowers which he incorporated into a society. This was a most important and far-seeing step which not only made possible the kind of research that he was planning for his own laboratory but was to have far-reaching consequences. A quarter of a century later physics laboratories all over the world had their Leiden-trained glassblowers without whose skill and competence they would have been lost. The main beneficiaries were, of course, the Dutch laboratories, and later the young electrical industry in Holland had reason to be grateful to the foresight which Kamerlingh Onnes had shown at the turn of the century.

Another instance of Onnes' long-range planning was the establishment of a scientific journal, devoted exclusively to the work of his laboratory. He ensured in this way that the cryogenic work which he was planning and which, as it actually turned out to be, should be unique in the world, was to be available to the scientific community in an unbroken and easily accessible form. With this, too, he scored a great success and for several decades to come the COMMUNICATIONS FROM THE PHYSICAL LABORATORY OF THE UNIVERSITY OF LEIDEN became the bible of low-temperature research.

The first major installation which Kamerlingh Onnes set up at Leiden bears already the clear stamp of his strategy of massive attack along the road to absolute zero. This was the large cascade-type liquefaction plant for oxygen, nitrogen and air which we have already mentioned in an earlier chapter. It was constructed in 1892–4 and on so large a scale that it satisfied the rapidly growing requirements of the Leiden laboratory for more than thirty years. It is also significant for Onnes that this large plant worked immediately with excellent efficiency; the result of ten years of careful planning.

The next fifteen years between this first achievement and the liquefaction of helium saw an ever-increasing volume of the most careful measurements, mostly connected with the equation of state and following in the footsteps of van der Waals. Gradually Leiden became a Mecca for scientific pilgrims. The spirit of Leiden was the opposite of the jealously-guarded sanctum which Dewar had created at the Royal Institution and to which only he and his close collaborators had access. Onnes threw the doors of the Leiden laboratory wide open to scientists from all over the world who wished to carry out cryogenic work. This liberal attitude of granting facilities, as well as the impressive scale on which they were provided at the Leiden laboratory, are good reasons why they remained unique for more than a quarter of a century.

It must not be thought that Onnes achieved his successes by careful scientific planning only. It required all the skill of a clever diplomatist to dig up the funds which he needed for buying equipment, paying for assistance and buildings and seeing the work of his laboratory in print. In addition, he had unforeseen troubles. In the late 'nineties when he began to design the hydrogen liquefaction, the whole work planned for his laboratory was threatened very seriously. Somebody with a deep sense of public responsibility, which was fully equalled by ignorance of technical matters – a combination which occurs not infrequently – petitioned the Ministry of the Interior to stop the work of the laboratory. It had become known, he pointed out, that the professor was experimenting with compressed gases and this might spell danger to persons and buildings. It should be forbidden. So work at Leiden had to come to a standstill until a government commission was appointed, had studied the problem, and given its verdict. The commission on which, among others, van der Waals served, wisely pointed out 'that the energy released by the bursting of a cylinder of compressed gas is much smaller than that liberated by the burning of 3 kilograms of gunpowder, an amount the possession and transportation of which is permitted without any difficulty'. Altogether, they were fully satisfied by the safety precautions used at the laboratory. Onnes had in the meantime asked experts abroad to testify on his behalf and Dewar wrote that it would be 'a terrible disaster for science in your country (and universal science) if limitations should be laid on your splendid cryogenic laboratory and the fine work you are doing'. So Onnes won his case and after two years was able to continue his researches.

Methodically, step by step Kamerlingh Onnes made his preparations for the assault on helium. Hydrogen was not liquefied in appreciable quantity in Leiden until 1906, eight years after Dewar's first liquefaction. However, the Leiden installation was designed to operate with the reliability of factory equipment and it produced up to four litres of liquid per hour. Now, with equipment at his disposal which was capable of producing without any hitch large quantities of liquid air and liquid hydrogen, Kamerlingh Onnes was in a far superior position to Dewar or Olszewski whose apparatuses were mere toys compared with the Leiden installation.

The supply of a sufficient quantity of helium gas of adequate purity, too, was assured by careful planning and the diplomatic use of a helpful connection. The gas was to be extracted from monazite sand, and Kamerlingh Onnes records that he was able to procure large quantities on favourable terms through 'the Office of Commercial Intelligence at Amsterdam under the direction of my brother'. He then goes on to say that after this 'the preparation of pure helium in large quantities became chiefly a matter of perseverence and care'. At the same time the design of the helium liquefier, based on the experience with the hydrogen machine, was completed and given over into the capable hands of Messrs. Flim and Kesselring, the heads of the mechanics and glassblowing workshops. Finally, in early June 1908 everything was set and ready.

An account of the memorable day is given in the famous Leiden Communication No. 108 which makes truly epic reading. In the introduction Onnes gives a short sketch of the history up to date. He recalls Dewar's first estimate of 5 to 6 K for the critical point of helium which was followed by Olszewski's depressing forecast of 2 K, a figure which was supported by Onnes' own earlier work at Leiden. However, he knew that the only reasonable thing was to wait for the results from careful measurements of the helium isotherms. These had at last been obtained during the previous year and they gave hope again. The forecast derived from the new Leiden measurements tallied with Dewar's. Now the road seemed clear enough to finalise the design of the helium liquefier and to set up the cycle in which the precious gas would be circulated from storage to the liquefier and back to storage without being lost. A description of the liquefier and the rest of the installation forms the next chapter.

Then follows the description of the experiment itself. On the 9th July no less than 75 litres of liquid air had been made in preparation for the final assault, which began at 5.45 a.m. on the 10th July with the liquefaction of hydrogen of which 20 litres were required. They were ready for use in the helium liquefier at 1.30 p.m. Extreme caution was now needed in pre-cooling this apparatus with liquid hydrogen. The tiniest quantity of atmospheric air which might inadvertently be introduced in one of the numerous operations ahead would jeopardise the final experiment. It would turn solid in the liquid hydrogen and frost up the glass of the helium container, making observation impossible. This difficult series of operations had been planned and rehearsed by Flim, who was in charge and succeeded in carrying them through without serious mishap. Helium circulation started at 4.20 p.m. and from then on the central cryostat of the liquefier set out into new regions of uncharted low temperatures. It contained a helium gas thermometer which was to give an indication of the way in which the experiment was progressing, and this was now the only guide.

For a long time the indicator hardly changed position and no appreciable cooling seemed to take place. Various settings of the expansion valve and adjustments of the gas pressure were tried until finally a gradual drop of the thermometer set in. The temperature of the central vessel seemed to fall slowly and fitfully and then the decrease stopped again entirely. Meanwhile the last of the available liquid hydrogen was being used and there was still no sign of helium liquefaction. At 7.30 p.m. it seemed that the attempt had failed.

In the course of the day word had gone around in the University that the great experiment was under way and some of Onnes' colleagues drifted in to see how things were going. It was at this critical moment when there seemed little hope of success that one of these visitors, Professor Schreinemakers, suggested that the steady refusal of the thermometer to fall further might be due to the fact that it was actually immersed in a boiling liquid. Perhaps helium had been liquefied after all but was difficult to see. What about illuminating the vessel from below? This was tried and suddenly the liquid level appeared, now clearly visible by the reflection of the light underneath. The central vessel was almost completely filled with liquid helium. Kamerlingh Onnes had fulfilled Lavoisier's prophecy. The last natural gas had been turned into a liquid.

The first flush of excitement had hardly subsided when another visitor came in to see the new liquid. This was Professor Kuenen and he pointed out that the aspect of liquid helium was very different from that of either air or hydrogen. In particular, he was astonished by the fact that the meniscus of the liquid was hardly visible where it touched the glass wall. He compared its appearance with that of carbon dioxide near the critical point. However, the real reason, which was understood only much later, is connected with those entirely new aspects of matter which only become apparent as absolute zero is approached.

These new aspects became manifest once more in the last part of the experiment recorded in Communication 108, but again their meaning remained hidden and their significance passed as yet unnoticed.

More than 60 cubic centimetres of liquid helium had been made in this first attempt and in the last phase of the experiment Onnes tried to obtain solid helium by letting the liquid boil under reduced pressure. In order to obtain the lowest possible temperature he allowed all the liquid to evaporate until only about ten cubic centimetres were left. Then he connected the helium cryostat to a strong pump which reduced the pressure above the liquid to one-hundredth of an atmosphere, but no sign of solid helium appeared. He concluded that the triple point of helium must lie below this limit of his experiment but he did not realize then that he had reached a temperature approaching closely only one degree above the absolute zero. Twice later in his researches was he puzzled by the failure of liquid helium to be cooled into a solid and he did not live to see the ultimate explanation of the enigma.

When Communication No. 108 was written, the great triumph was the actual liquefaction of helium itself, the last step towards the absolute zero which seemed possible at that time. The author was as yet unaware that, beyond reaching a new low temperature, he had opened up a new world of strange phenomena which were to affect profoundly the concepts of physics. The great experiment had been successful. At about 10 p.m. the work was stopped and

Onnes wrote: 'Not only had the apparatus been strained to the uttermost during this experiment and its preparation, but the utmost had also been demanded from my assistants.'

The paper ends with a summary of the newly observed properties of liquid helium. Kamerlingh Onnes prefaces these with a chivalrous aside about his defeated competitor, Dewar, by pointing out how correct many of his predictions were. However, other phenomena were observed which were quite unexpected. In addition to the very small surface tension which made him fail to see the meniscus at first, there was the failure of the liquid to solidify. Another property which astonished him, and which was to assume great importance later, was the low density of the liquid. Liquid helium turned out to be about eight times lighter than water and this was much less than had been expected.

Kamerlingh Onnes returned to the assault on the lowest temperatures in the following year, and again a year later, in 1910. In 1909 he was able to reduce the pressure of vapour above the liquid to 2 mm of mercury, which corresponds to 1·38 K, and still more powerful pumps used in the next attempt reduced the vapour pressure to 0·2 mm, yielding 1·04 K. Still, helium remained a liquid, even at these low temperatures. The limit had been reached to which the best pumps available at the time were able to reduce the pressure, and for a significant advance over his last achievement Onnes had to wait for more than ten years. By then a new type of pump had been developed in response to the demand for highly evacuated lamp bulbs and radio valves.

However, the general interest of Onnes and his co-workers was deflected from the attainment of very low temperatures by something which happened in 1911 and which changed the trend of research work in Leiden. With a new temperature range, that of liquid helium, now at his disposal, Kamerlingh Onnes turned to investigations of the properties of matter within a few degrees of the absolute zero. To some extent the choice of experiments which were to be carried out depended on the technical difficulties which would be involved. A measurement which can be made at any

4·1 Whereas in a normal metal the electrical resistivity becomes constant at low temperatures, the resistance of a superconductor vanishes suddenly at the transition point.

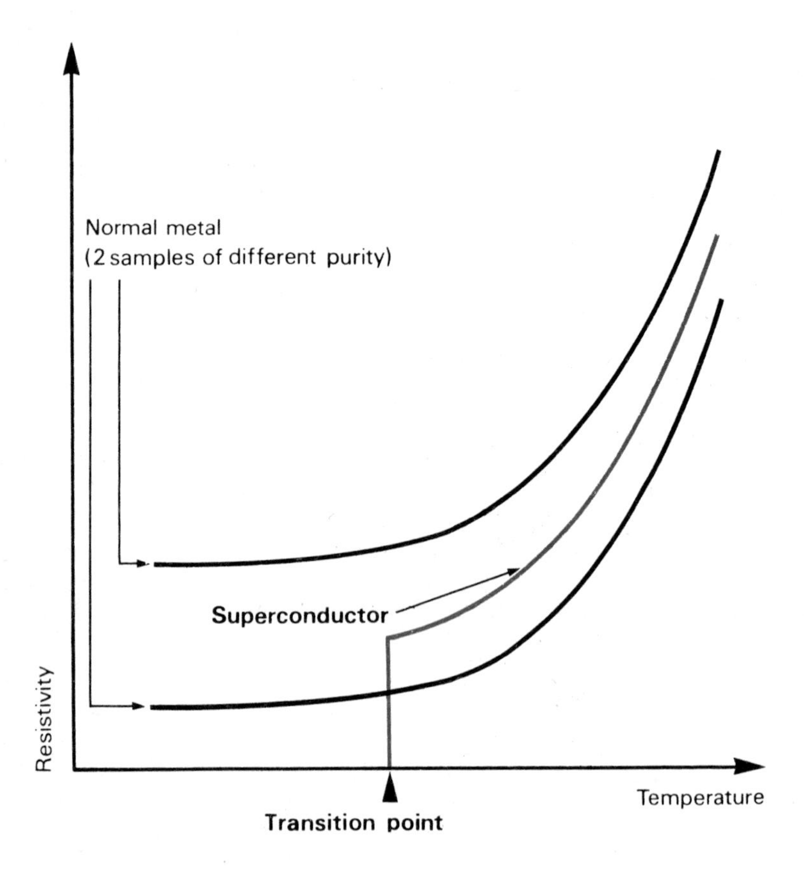

temperature with comparative ease is the determination of the electrical resistance of a wire. In addition, the question of the electrical resistivity of pure metals had by then achieved considerable importance. The work of Nernst in Berlin, which will be discussed in detail in a later chapter, suggested that, with a lowering of the temperature, the resistance of a pure metal should become gradually smaller and eventually disappear completely at absolute zero. The same question had been investigated by Dewar at the temperature of liquid hydrogen and his results, far from improving an understanding of the problem, had made it less clear. Dewar had found that the resistivity of platinum was dropping at a lower rate than expected as the temperature decreased. This seemed to support another theory according to which the electrons which carry the current should at absolute zero become fixed to the atoms. This, of course, means that the electrical resistance would become infinitely large at the lowest temperatures. Dewar's observations on platinum were therefore interpreted as the onset of this phenomenon in which the resistivity was expected to pass through a minimum somewhere below the hydrogen range.

Kamerlingh Onnes took up the research where Dewar had to leave it and investigated the resistivity of platinum at helium temperatures. The results were disappointing in that they neither confirmed nor contradicted either theory. The resistance neither fell nor rose as the temperature was lowered but remained quite steady. Onnes, however, picked up the right clue and interpreted it correctly. He noticed that the absolute value of the temperature independent resistance varied from sample to sample and that it was lower the purer the metal was (figure 4·1). From this he concluded that Nernst was probably right; the resistance should become vanishingly small as absolute zero was approached but that it was prevented from doing so by the presence of impurities.

The next task was therefore to reduce as far as possible this 'residual resistance' by using still purer metal specimens. Since Onnes knew that gold can be refined much better than platinum, he changed over to the investigation of the purest gold wires which he could obtain. This step proved successful in that much lower values of the resistivity were recorded than with platinum. Still, the same pattern was repeated; the resistance fell with increasing purity. Onnes was now certain that he was on the right path. He had his own ideas as to the behaviour of electrical resistivity at the lowest temperatures which turned out to be quite wrong and, as a result, he failed at first to see the magnitude of the discovery he was about to make.

Dewar's and his own measurements at higher temperatures had shown for all metals a fall in resistivity which, when extrapolated to very low temperature, indicated that the resistance would completely vanish a few degrees above the absolute zero. He had correctly interpreted the levelling off of the resistance curve of platinum as due to impurities and was now looking for a metal which he could get into a sufficiently pure state to show that its resistance actually vanished at temperatures which he could reach with liquid helium. This was rather different from the predictions of Nernst and his school who expected zero resistivity to occur only at absolute zero. However, based on his results at higher temperatures and on a theory which proved to be quite incorrect, Onnes proposed a formula according to which the resistance of pure metals should gradually drop to zero at helium temperatures.

There was one other metal which might be obtained in an even purer state than gold, and that was mercury. Being a liquid at room temperatures, it can be distilled and re-distilled again and again until an extreme degree of purity is reached. The results were communicated to the Netherlands Royal Academy on the 28th April 1911, when Onnes reported that mercury, as well as a sample of very pure gold, had, at helium temperature, reached resistivities so low that his instruments had failed to detect them. He was particularly intrigued with the behaviour of the mercury sample because it had still a fairly high resistance at liquid hydrogen temperatures and could also be recorded at the boiling point of liquid helium but then vanished at lower temperatures. Kamerlingh Onnes was cautiously triumphant, pointing out that this was the result which he had expected on the basis of his formula. Nevertheless, true to form he ends on a note of warning that these were only preliminary results and that more accurate measurements were required.

The next report was made only one month later, on the 27th May. The accuracy of the measurements had now been increased but the result obtained was somewhat unexpected. The resistivity of mercury, far from vanishing gradually, seemed to drop sharply to an immeasurably small value just below the boiling point of helium (figure 4·1). Onnes notes that this does not agree with his formula. While this communication was headed: 'The disappearance of the resistance of mercury', the next one, seven months later, significantly bears the title: 'On the sudden change in the rate at which the resistance of mercury disappears.' Like the two previous papers this, too, is a very short one. It confirms the earlier result of a rapid disappearance of the resistance which had now been narrowed down to within only two hundredths of a degree.

Then over a year passed by without further reference to this work. However, a spate of papers published in 1913 makes it clear that this was not due to inactivity. On the contrary, work on the peculiar behaviour of mercury seems to have gone on all through 1912 at a high rate. A vague indication of the trouble into which Onnes was now running had been given in the closing sentence of the last communication in 1911. He referred to attempts at passing high currents through his mercury sample but added that the phenomena which then occur are peculiar. The four papers published between February and May 1913 allow us to reconstruct to some extent the course of events in the intervening period. Every new experiment not only confirmed the result that the resistance of mercury completely disappeared just below the boiling point of helium but it also showed that this was not the phenomenon which Onnes had predicted. Moreover, gold is no longer mentioned in any of the papers and we must assume either that work on it was discontinued or that Onnes had realised that, contrary to his first communication, gold behaves quite differently from mercury. This was perhaps one of the reasons why he was reluctant to report on his work for the time being, since it was now quite possible that the behaviour of mercury was peculiar to this metal. In any case, the publication of his four papers followed the discovery, in December 1912, that tin and lead also lose their resistivity.

In the second of the 1913 papers, published in March, there

appears for the first time the word *superconductivity*, the term by which Onnes' great discovery was to become known. He appears to have introduced it merely as a space-saving expression and its first use does not, as yet, seem to indicate a full understanding of the new effect. Indeed, when lecturing in September at a scientific meeting in Washington, he referred to his work on mercury as realising 'the superconductive state, the existence of which had only become probable by the experiments with gold and platinum'. At this stage superconductivity must still have appeared to Onnes as something which, though remarkable, was not totally unexpected. He evidently regarded it as an extreme case of the usual mechanism by which electricity is conducted through a metal.

We cannot at this stage enter into the peculiarities which Kamerlingh Onnes observed in his early experiments on superconductive mercury and which made him hesitate to publish his observations. The pattern which slowly emerged from his own observations and those of others made it clear that superconductivity, far from being an extreme case of ordinary electrical conduction, is a phenomenon which has little connection with any of the known properties of matter. It defied theoretical interpretation for the better part of half a century and even today our understanding of it is far from complete. We will return to it in a later chapter.

Superconductivity provides an object lesson of how difficult it can be to make a great discovery, even when the effect is striking and requires no more than a fairly simple experiment. The scientist, who regards the happenings which he observes as evidence of an organic fabric, will always try to link up what he finds with the established body of knowledge. This makes it difficult for him to comprehend at first sight a new phenomenon which has no connection with known fact. It was quite inevitable that Kamerlingh Onnes should try to find a place for superconductivity within the framework of ordinary conduction. In his particular case the issue was further confused by his own – completely erroneous – prediction that ordinary conduction should become infinite at some low temperature. Only gradually, as more and more of his experiments demonstrated how profoundly superconductivity departs from the established pattern of electrical conduction, did Kamerlingh Onnes see the magnitude of his discovery.

More than twenty years later, at the end of his life, some very essential features of superconductivity remained still undiscovered. Even so, Onnes had the satisfaction of knowing that in superconductivity he had discovered an entirely new aspect of matter which until then had lain hidden in the world of very low temperatures that he had opened up. Of another, equally strange, phenomenon he had a few glimpses without, however, grasping its fundamental significance. It was much the same story as with superconductivity; the very magnitude of the impending discovery prevented his making it.

Here the substance on which the discovery was to be made was even nearer at hand. It was liquid helium itself. It will be remembered that on the day of first liquefaction Onnes had made a rough estimate of the density and had found it surprisingly low. He decided to investigate whether it varied with temperature and noted to his surprise that, apart from having a generally low value, the density of liquid helium passes through a maximum at about 2·2 K. He published this result early in 1911 but the discovery of superconductivity in the following months now claimed most of his attention. Late in 1913 he referred to this strange phenomenon in his Nobel Laureate Lecture, intimating that, like the new electrical effects, it might be connected with the quantisation of energy discovered by Planck at the turn of the century. Kamerlingh Onnes finally returned to the problem in 1924 when he published a set of careful measurements, made together with Boks, which extended down to just above 1 K. The density curve (figure 10·2) obtained by them showed a rise from the boiling point down to 2·2 K, following a gradual decline below this temperature. Thus there could be no doubt that, as the temperature is lowered, liquid helium contracts, as one would have expected, but it then starts again to expand towards absolute zero. A similar behaviour is, admittedly, observed for water at 4°C but the water molecule is of a more complex structure for which anomalous behaviour is not too astonishing. However, the helium atom is the most symmetrical of any found in nature and this made the density maximum quite enigmatic.

Onnes fully realised the importance of his observation and planned a full-scale attack. He was aided in it by an American guest worker, Leo Dana, and they first set out to measure the latent heat of evaporation which is required to turn the liquid into vapour. Following this quantity from the boiling point down to lower temperatures, they found indeed a dip in the curve but it was small and only just outside the accuracy of their measurement. At the same time they had embarked on determinations of the specific heat and here they were confronted with an effect of such magnitude that they did not believe it. Their published data only extend down to 2·5 K and show an unexciting result; a curve which gently decreases to low temperatures. This is the kind of behaviour which could be expected. Their work had, in fact, been extended down to the temperature where the density maximum occurs. Here the results had been quite extraordinary, yielding varying and very high values. Onnes and Dana felt that these odd results could not be true but were probably caused by a fault in their apparatus. Accordingly, they decided not to include them in their paper which was published in 1926. When it appeared in print, Dana had returned to America and Onnes was dead. The work on the strange behaviour of liquid helium, to which we shall return in a later chapter, had to be continued by his successor and by research in Oxford, Cambridge and Moscow. For, by then, the Leiden laboratory had lost its monopoly position.

With the death of Onnes an era of low temperature research came to an end. The great laboratory at Leiden which now bears his name has continued its work on a magnificent scale to this day. However, an ever increasing number of other research establishments have, for the past fifty years, shared in the investigations close to absolute zero. They all owe a great debt to Kamerlingh Onnes, whose pioneering not only opened up the new range of liquid helium temperatures but set a new standard in planning research and directing a laboratory. He was an outstanding scientist whose discovery of new phenomena earned him the Nobel Prize and a host of other honours. His greatest achievement, however, is the originality with which he created the laboratory that was to become the pattern for research in the twentieth century. Unlike Dewar, he was a good diplomatist who knew how to get on well with people although, at times, he expected them to work to the utmost of their capacity. In all his achievements he gave credit to those whose work had made his success possible, and they remained devoted to him. A significant story is told about his funeral. The cortege was moving from the church in town to the cemetery, with the 'meesters' Flim, the head of the workshop, and Kesselring, the chief glassblower, walking behind the hearse. It seems that the service had lasted somewhat longer than anticipated and the hearse had to hurry to reach the churchyard in time. Said Flim to Kesselring: 'That's the old man all right. Even now he makes us run.'

As the Kamerlingh Onnes era of liquid helium was drawing to a close, a new age dawned. In his last effort to reach a very low temperature, in 1922, Kamerlingh Onnes had used a battery of twelve of the new diffusion pumps, which had been developed by Langmuir, to pump off the vapour over liquid helium in an elaborately shielded cryostat. With this colossal armoury he reached 0·83 K. The communication was given as an address to the Faraday Society and is significantly called: 'On the lowest temperature yet obtained.' At the end he asks the question whether this had to be considered as the final step towards absolute zero. He points out that, unless another substance – more volatile than helium – can be discovered, the absolute limit set towards low temperatures had been reached. He rejects this argument with the words: 'We cannot accept such a limit otherwise than as a provisional one', and then concludes: 'We may feel sure that the difficulty which has now arisen in our way will be overcome also and that the first thing needed is long and patient investigation of the properties of matter at the lowest temperatures we can reach.'

This prophecy came true less than a year after it had been made. The paper which contained the observations that were to open up the new temperature range below 1 K was published by himself and H. R. Woltjer as Communication No. 167c. He held in his hand the key to temperatures ten and a hundred times lower than those which he had just obtained – but he did not recognise its significance. A few years later others who had grown up in a different school of thought saw the meaning of the key which Onnes himself had provided.

The new ideas from which progress to lower temperatures was to spring had been formed when Kamerlingh Onnes was well on in his fifties. He was aware of them but to him, who had grown up in the firm and authoritative teaching of classical physics, they remained alien. They are disturbing revolutionary ideas which have changed and are still changing our accepted concepts. At the same time the demolition of the well-established pattern of physical laws has opened new and exciting vistas. With Kamerlingh Onnes we must leave the old physics with its security and its limitations, sound and confined as the Victorian age. However, before we can continue our journey towards the absolute zero, we must make a long detour through the development of modern physics.

5 The third law

The first signs and portents of new and unexpected laws of physics made their appearance at Dewar's liquefaction of hydrogen. The heat required to evaporate the liquid was even smaller than was to be expected from Trouton's rule. The thermo-element had ceased to be an indicator of temperature and the electrical resistance thermometer had yielded readings which were far too high. There were also the strangely low values of the specific heats which soon would become the most important clue in the search for new ideas. With each new experiment more and more evidence accumulated that in the vicinity of absolute zero the accepted concepts failed to be useful. As the new territory of very low temperatures began to be explored and charted, the familiar landscape of the physical pattern changed. It was as yet too early to understand the cause and meaning of this change but of its existence there could be no doubt.

Looking back at the discoveries of facts and concepts which soon were to shake the proud edifice of 'classical' physics, we see them arise completely unconnected and in a form which did not suggest any relation between them. At the same time as Dewar lowered the energy of matter by cooling it down to twenty degrees above absolute zero, Max Planck in Berlin had to conclude, much against his liking, that energy is not the homogeneous fluid which it had always been assumed to be but that it has structure. Walther Nernst in Goettingen tried to find rhyme and reason for the processes by which a new chemical industry manufactured its products, and Einstein, a young man working at the Patent Office in Berne, began to form the ideas which were to link up these isolated new trends.

From the pioneering work of the experimentalists in France, Poland, England and Holland the German theoreticians were to build the edifice of new concepts of 'quantum' physics which was needed to explain the phenomena discovered at low temperatures. Nernst, Planck and Einstein were its chief architects. Any public opinion poll on their greatness would list their names in the reverse order but as far as the understanding of the world near absolute zero is concerned we have to abide by our sequence.

Walther Nernst was born in 1864 as the son of a Prussian country judge in Briesen, a small town close to the border of Tzarist Russia. He was a short man who grew bald early, originally wanted to become a poet, and loved the theatre. He played the life-long act of a little, rather innocent and often mildly astonished man of simple sincerity, but behind this was hidden one of the most versatile and penetrating minds and a sarcastic sense of humour. His famous book, *Theoretical Chemistry*, ushered in a new era of research and set the pattern of thinking for a whole generation. This was the generation of men who set out to transform the approach to chemistry from the empirical laboratory stage to that of the large industrial factory, relying for its operation on accurate prediction. The title of the book summarises Nernst's life work.

When, in 1871, Bismarck created the Hohenzollern Empire, the new Reich came late into the company of great powers. Such colonies as it had yielded little except prestige and trouble, and the Fatherland itself was singularly poor in minerals except coal. In particular, there were no nitrates, but nitrogen in the form of chemical compounds is a basic necessity, for fertilisers in peace and for explosives in war. The new Reich might need them for either purpose. On the other hand, nitrogen as an element is available everywhere in abundance since it forms 80 per cent of the atmospheric air. Hydrogen is equally abundant and the two elements combine to form ammonia. Once ammonia is obtained, the rest is easy. Thus the problem with which the German chemists were faced was to find a way of combining the gases nitrogen and hydrogen, but for its solution the old alchemistic hit-or-miss approach was not suitable. In any case, it had been tried and had not yielded any success.

It was this problem in its most general form which occupied Nernst. When he had come to Goettingen, the University had built for him the first laboratory for physical chemistry. This was a new subject which had arisen in the preceding decades through the application of physical methods, in both experiment and thought, to chemical problems. Its main line of approach was the use of thermodynamics, the conceptual framework of which had originally been developed in relation to heat engines. Thermodynamics had soon outgrown these somewhat narrow confines and had proved itself to be a most powerful tool in the solution of any problem involving changes in temperature.

Before we can continue our story of low-temperature physics, we must pause to take another look at the basic laws of thermodynamics and their operation. They form the first step in the understanding of the new methods which have been developed to reach temperatures much lower than those of liquid helium and also of the strange phenomena encountered in the approach to absolute zero.

Thermodynamics had profited greatly from the labours of one Benjamin Thomson, a man of a lively and adventurous frame of mind, born in 1753 at Woburn, Massachusetts, who in turn became Count Rumford of Bavaria and one of the founders of the Royal Institution in London. After Lavoisier had been guillotined in the French Revolution in 1794, Rumford's ambition drove him to the unwise step of marrying the great scientist's widow. Alas, he was no Lavoisier who had been able to dominate her into sitting obediently beside him during experiments, taking notes. She led the poor Count a hell of a life.

While boring cannon for the Duke of Bavaria, Rumford noted that the metal was getting hot – in fact it got hotter the blunter the boring tool was. Also, the cannon was the hottest place in his 'experiment', and no combustion which might produce heat was taking place anywhere. Rumford rightly concluded that heat was created by friction and that in this process of heat generation mechanical work, the rotary motion of the boring drill, was consumed. Mechanical work was being transformed into heat. This is the opposite process from that taking place in a steam engine, where the heat generated by combustion in the boiler is being changed into work; the rotary motion of the driving shaft.

This and other observations of a similar kind gradually led to the understanding that there exists a physical quantity which can appear in different aspects but, since it can be transformed from one to the other, it must still be the same thing. This quantity is the energy and it was only one step further to conclude that its transformation will always take place without loss. The statement that energy cannot be created or destroyed became known as the law of conservation of energy or, as it is sometimes called, the first law of thermodynamics.

The concept of energy was, of course, well known since Newton's time because it arises from the motion of a body of mass m and travelling with the velocity v as $\frac{1}{2}mv^2$. In Newtonian mechanics it is, however, not as obviously important as another combination of mass and velocity, mv, the so-called momentum, the conservation of which Newton had already postulated. Both energy and momentum, in the way in which we have written them down here, refer explicitly to the motion of a body and do not lend themselves to the interpretation of other phenomena such as an electric current or, for that matter, to the boring of a cannon. Indeed, while the conservation of momentum is now recognised as one of the basic principles of particle physics, its applicability has remained strictly confined to that field. The energy, on the other hand, as was shown by Rumford, can appear in the form of heat or, as was later discovered, in the form of an electric current or of a chemical reaction. A law describing the behaviour of energy is therefore bound to be of great general significance. The discovery that energy is always conserved immediately puts an important limitation on the processes which can occur in nature; only those are permitted in which the energy remains unchanged.

Such limiting laws, of which the first law of thermodynamics is possibly the most important one, are the lifeblood of all science. The electricity company sells us energy in the form of electric current and, thanks to the conservation of energy, they can guarantee that exactly the same amount of energy for which they charge will make its appearance in the form of heat or light in our home. Out of all the possible quantities of heat which might conceivably have been produced by the quantity of current which we have consumed, the law of conservation of energy has selected that quantity which has the same energy. The guarantee of the electricity company is just one instance of the accurate prediction about the course of natural phenomena which this law allows us to make.

Proper understanding of the conservation of energy was complicated by the work on heat engines which, as we mentioned in chapter 1, was begun by Carnot. This complication is produced by the fact that there is something unsymmetrical about the transformations of energy. While, for instance, in experiments similar to the boring of cannon *all* mechanical energy can be transformed without residue into heat, the opposite is never the case. When heat is fed into a steam engine, only part of this energy can be made to turn the shaft and an unavoidable residue will be discarded in the form of waste heat in the condenser of the engine. We have chosen purposely the word 'unavoidable' because it indicates the existence of another of these limiting laws: the second law of thermodynamics.

It is significant that, while the first law is always stated with great clarity in even the most elementary textbooks of physics, the second one is either omitted or surrounded with such waffle as will ensure that the reader is not likely to be clearer about it than the author was. In short, it is considered 'difficult', which is a pity since it is a beautiful and satisfying theorem. Admittedly, its thermodynamic formulation, while perfectly clear and unambiguous, does not present a concept which can be grasped on the basis of everyday experience such as temperature or pressure but, as we shall see, the kinetic picture removes this difficulty.

After Carnot the next step in understanding the transformation of heat into other forms of energy was taken by a German, Rudolf Emanuel Clausius. He turned his attention to a distinction between that fraction of heat which in such a process can appear as mechanical energy and that which must be discarded as waste heat. He called the first the 'free' energy, and described the second by a new term, the 'entropy'. It is this entropy which is required for the

5·1 Red and white sand will mix when stirred clockwise but not unmix when stirred anti-clockwise.

statement of the second law of thermodynamics. This law simply says that only those processes can take place in which the entropy increases or, at best, remains constant. In other words, the second law of thermodynamics excludes all processes in which the entropy decreases.

Any proper understanding of this important law, of course, requires that one can form some concrete idea of the concept of entropy. Thermodynamics is not very helpful here. It just tells us that entropy is a quantity of heat divided by the absolute temperature. This is a most useful, and in fact simple, term to be put into any equation but it fails to convey much meaning. This difficulty remained until, at the end of the nineteenth century, the formalism of thermodynamics was combined with the concepts derived from the kinetic theory of heat into an exceedingly powerful method of theoretical interpretation which is known as statistical thermodynamics. The introduction of statistical methods, mainly by the

Viennese Ludwig Boltzmann, revealed the true nature of entropy. Entropy turned out to be the degree of *disorder* of the system.

Suddenly not only the meaning of entropy became clear but the enigmatic second law of thermodynamics turned out to be a well-known principle of everyday experience. Things which are in an orderly state when we start, say, the books on the shelves of a library, tend to get into disorder when the library is in use. For a more concrete example we can take a small glass jar and fill it half with white and half with red sand (figure 5·1). We now take a spoon and stir the contents of the jar a hundred times in a clockwise direction and the result will be pink sand. The orderly pattern of separate red and white sand has been destroyed. Stirring has increased the entropy. One might object to this conclusion by saying that we have not yet proved the law which says the entropy must increase in *any* process because we have simply happened to pick out a particular example in which it does so. So let us now reverse our process by stirring the sand a hundred times in an anti-clockwise direction. The result will, of course, not be a separation into the original orderly pattern of white and red sand but our sand will get even more thoroughly mixed and become, if anything, pinker. In other words, the entropy has increased further, just as is required by the second law. What is most impressive is that we do not even have to do this little experiment since we are already fully convinced of its outcome beforehand. The conviction stems from everyday experience and it is pleasing to realise that the true meaning of this 'difficult' second law of thermodynamics is, in fact, so deeply ingrained in the commonplace pattern of our life that we do not have to stop and think about it.

Now we can understand the strange lack of symmetry in the transformation between mechanical energy and heat. A moving train is brought to a standstill by friction of the brakeshoes against the wheel. In this process the shoes and wheel get hot; the energy of motion of the train has been transformed into heat. On the other hand, if we heat up the wheel of a stationary train, we will not set it into motion. Here again the kinetic picture will make

5·2 On applying the brake the ordered motion of atoms in the train wheel is transformed into disordered motion.

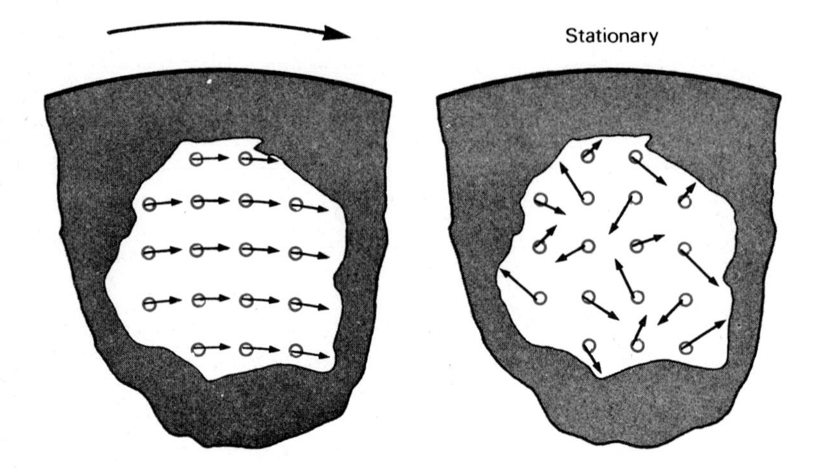

things clear. The wheel, and all the iron atoms of which it consists are moving in the same direction. If we make for ourselves a microscopic picture of any part of the wheel in which we can see the individual atoms, we can indicate the motion of these atoms by little arrows (figure 5·2). Since they all take part in the motion of the wheel in the same manner, the little arrows will all have equal length and direction. These arrows now indicate the velocities of each atom and together they make up the kinetic energy of the wheel as a whole. When the train is stopped by friction in the brake, this energy is transformed into heat. On the atomic scale this simply means increased irregular motion of the atoms and in our picture this will be shown by a redistribution of the length and direction of the arrows. The sum total of the arrows is still the same but now their lengths and directions are distributed at random. The orderly pattern of mechanical energy has been changed to the disorderly pattern of heat motion. The entropy has increased.

Again, as in the case of the jar filled with sand, little imagination is required to see that the process cannot be reversed, at least not completely, and not by simply heating up the wheel. From the heat which we have just offered to provide to the wheel, a small amount of orderliness can still be extracted but this is only a small fraction of the energy that originally had been turned into heat on braking. If we want to use heat to make the train start up again, we have to go through the complicated mechanism of a steam engine. Even then, as mentioned earlier, the heat energy passing through it will have to be much larger than the mechanical energy which we shall be able to extract. In letting the heat energy pass from the high temperature of the boiler to the lower temperature of the condenser through the mechanism of the engine, we can abstract as a small fraction the mechanical energy to run the train. This was possible because heat at the high temperature is somewhat less disorderly than that at the lower temperature. The total entropy has, of course, increased throughout the whole operation, as is demanded by the second law of thermodynamics.

Our example illustrates that, since heat energy is random motion of the atoms, it cannot be transformed into mechanical energy without a residue because no process can take place in any system which would allow the overall disorder to decrease. Taking the example of the jar with the white and red sand or that of the rotation of the wheel turned into random motion of the iron atoms, we can see the growth of disorder fairly clearly. In the operation of the steam engine it is not nearly so apparent and there are many cases where the instructive statistical pattern cannot be recognised at all. What are we to do then? How can we use the law of entropy increase if we are unable to see any change in the state of orderliness? Here we are saved by turning back to thermodynamics. Its definition that entropy is a quantity of heat divided by the absolute temperature at which this heat is developed may not convey to us the meaning of entropy but it allows us to measure entropy. Once a physical quantity can be accurately determined by measurement, we can use it with confidence.

At last our necessary excursion into the basic laws of thermodynamics is finished and we can return to the problem which faced the physico-chemists at the turn of the century and which was solved by Nernst. The problem, let us recall, was the prediction of chemical reactions. The great French chemist Berthelot thought in the middle of the nineteenth century that he had found the solution. He postulated that any reaction would always proceed in the direction of maximum heat production. This turned out to be wrong, but not very wrong. Most processes which take place spontaneously do indeed produce heat, like the burning of coal or the release of atomic energy. However, there are a few which spontaneously produce cold, such as the melting of ice when salt is poured onto it. The object of this procedure is the unfreezing of railway points and, while it serves its purpose in melting the ice, the solution of salt in water with which one ends up is, in fact, colder than the ice was originally. It is now easy to see where Berthelot was wrong. It is not the heat that tends to a maximum but the entropy, that means the degree of disorder. Admittedly, we have lost some disorder in the reaction of ice with salt by the fact that we lost heat. However, this is more than balanced by the increase in disorder due to the mixing up of water with salt, and so the overall disorder has grown as is required by the second law of thermodynamics. Just as in the case with sand of different colours, water and salt will mix, but not unmix, spontaneously.

It is not the total energy which, as Berthelot thought, governs the direction of the process but the free energy since the entropy determines the difference between the two. This important fact was recognised in 1883 by the Dutchman, Jacobus Hendricus van't Hoff, who in this way gave a clear definition to the old alchemist notion of 'affinity' between substances. An equation connecting the total and free energies with the absolute temperature had been derived a few years earlier from a combination of the first and second laws of thermodynamics by the great American physico-chemist, Josiah Willard Gibbs, and the German, Hermann von Helmholtz. If for the partners in the chemical reaction, say, hydrogen, nitrogen and ammonia, the free energies are known as functions of temperature, the chemical equilibrium can be predicted.

The total energy of a substance can be obtained by measuring its specific heat. Unfortunately, however, the mathematical form of the Gibbs–Helmholtz equation does not allow us to calculate the free energy from the total energy without some further assumption. This assumption was made by Nernst, who based his ingenious guess on the fact that Berthelot was almost right. This must mean that at normal temperatures the total and free energies cannot be too different and Nernst therefore postulated that, as absolute zero is approached, they will become equal.

More than half a century has passed since Nernst stated his postulate in 1906 and a large amount of experimental evidence gathered since then has amply proved that his guess was correct. Only a decade after Nernst's first publication of his theorem it had become acknowledged as the third law of thermodynamics.

The third law soon proved its worth in the prediction of chemical equilibria and when, eight years later, war broke out, synthetic nitrates were beginning to flow from the huge Leuna works into the German armament industry. Before that, however, the practical importance of the third law was already overshadowed by the attention which it focused on the approach to absolute zero. Its significance for low temperature physics is shown by the two further formulations in which Nernst's theorem can be expressed. One says that, while absolute zero can be approached to an arbitrary degree, it can never be reached, and the other says that at absolute zero the entropy becomes zero.

Although the first of these two statements seems the more dramatic one, it is far less important than the second. Since the time of Cailletet, when the attainment of low temperatures became an experimental possibility, none of those engaged in this work made reference to the possibility of actually attaining absolute zero. In connection with the liquefaction of hydrogen and later of helium, the approach to absolute zero is frequently mentioned, but it was taken for granted that eventually, at low enough temperatures, these substances, too, would freeze into a solid and become useless for further cooling. It seems that even before Nernst the experimentalists had accepted that their cooling processes would give out before absolute zero was reached, without requiring rigorous proof of this. There appears to be no record in the scientific literature before 1906 of anyone seriously considering the attainment of absolute zero by discussing the theoretical possibilities. If it had been done, the outcome would have been not very different. In retrospect we must conclude that anyone rigorously investigating the problem in, say, 1900, would have decided that absolute zero cannot be reached by a finite number of cooling stages although the reason for this failure is completely different from that given by the third law. There may have been a short twilight period of about five years before Nernst when 'classical' principles began to appear doubtful but quantum effects were as yet not properly recognised, when hopes could have been entertained of reaching absolute zero. Since, again, nobody seems to have bothered to raise such hopes, this phase had no effect on the subsequent development of ideas.

While the formulation of Nernst's theorem as the law of unattainability of absolute zero has gained currency in textbooks because it is a clear statement which is easy to remember, it cannot have made much difference to the efforts of men like Dewar or Kamerlingh Onnes. It does, however, affect in some measure the significance of absolute zero as the bottom of our temperature scale. This scale has no ceiling since we tacitly assume that, whenever we measure a very high temperature, a still higher one will exist and can possibly be attained. It is different with absolute zero because no temperature lower than it is conceivable. In this context the statement of the third law that beyond each low temperature which we produce there is a still lower one which we can attain without reaching absolute zero, must appear unsatisfactory. Moreover, with each further approach to absolute zero the interval separating us from this unattainable point decreases and must eventually appear as a ridiculously small fraction of a degree.

There are good reasons for thinking, instead of an additive scale of degrees of temperatures, in ratios. As a matter of fact it is in such ratios that temperature is defined by the second law of thermodynamics. It simply means that instead of counting up the number of degrees between, say, 1 K and 10 K which is, of course, nine or between 10 K and 100 K which is ninety, we look upon 10 K as ten times warmer than 1 K and on 100 K as ten times warmer than 10 K. Using ratios, we thus allot the same significance to the interval between 1 K and 10 K as to that between 10 K and 100 K, or to that between 100 K and 1,000 K, and so on. In this logarithmic scale we deal with powers of ten and write for 100 K = 10^2 K; for 1,000 K = 10^3 K and for a million degrees Kelvin 10^6 K. Equally, 1 K now becomes 10^0 K, 0·1 K becomes 10^{-1} K and a millionth of a degree absolute is 10^{-6} K. The absolute zero accordingly moves to minus infinity ($10^{-\infty}$ K) which seems most fitting for a temperature which can never be reached. Except for the different form of counting and a more convenient way of looking at absolute zero as infinitely distant, the logarithmic temperature scale is not a revolutionary step. The physical significance of the temperature concept remains quite unchanged.

While unattainability of absolute zero is not very significant for our understanding of the physical world, the other formulation of the third law which tells us that the entropy must become zero is a statement of fundamental importance. Its implications which were only gradually realised constitute a complete break with the physical concepts of the nineteenth century. Moreover, the disappearance of all disorder as absolute zero is approached turned out to be the key for the unusual phenomena first observed by Dewar and Kamerlingh Onnes.

The first reaction to Nernst's theorem was one of surprise that it was the entropy and not the energy which tends to zero. Until then the concept of absolute zero as a state of complete rest, first formulated by Amontons, had been taken for granted. In particular, since the kinetic theory explains temperature as the mean energy of molecular motion, it appeared inescapable that at zero temperature this motion must cease entirely. Now it turned out that, according to the third law of thermodynamics, some of the energy was retained even at absolute zero and the nature of this 'zero point energy' remained enigmatic and incomprehensible for a long time. Eventually it turned out to be a direct manifestation of the basic principle of quantum mechanics.

In addition to a number of new and unusual phenomena which arise at low temperatures as the result of growth in statistical order, the decrease in entropy also made the conventional physical pattern appear in a new light. An example is the well-known sequence of the three states of aggregation through which a substance will pass on cooling. As a gas is cooled it will first condense into the liquid state and on further cooling become a solid crystal. So far one had looked upon these changes as a direct consequence of the fact that temperature is the measure of the kinetic energy of atoms or molecules. As this energy diminishes, it can be balanced by the cohesive forces between the atoms and, instead of recoiling at each collision, the atoms will adhere to each other, first loosely in the liquid state, and eventually rigidly in the solid. It is, of course, the same sequence of events which we have mentioned earlier, in the discussion of the van der Waals equation. However, this sequence holds another aspect which we have so far neglected. In the gaseous state the atoms or molecules fill the whole of the available space in random motion (figure 5·3). In the liquid phase they still move about but now the substance confines itself to its own size, occupying only part of the available volume. Finally in the solid crystal the atoms are fixed into their positions in a regular array, called the lattice.

Thus, in the cooling of a substance the decrease in kinetic energy is not the whole story. Linked with it something else has happened; the pattern has become progressively more ordered. The entropy has decreased and, in a way, we can look upon this well-known sequence of gas, liquid, solid and upon the familiar processes of condensation and freezing as a manifestation of the third law of thermodynamics.

5·3 As a gas is liquefied and the liquid is frozen, the array of atoms becomes progressively more regular.

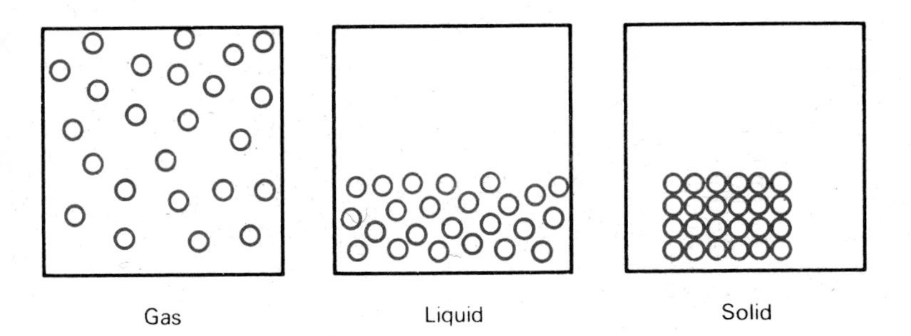

While in this way the states of aggregation acquire a new significance when we look upon the changes in orderliness which they represent, there exist other phenomena, also due to the third law, that are not so immediately apparent and therefore more astonishing. Since the wavelengths of x-rays are of the same order of magnitude as the spacing of atoms in a crystal lattice, x-rays have become a very powerful tool for the detailed investigation of these lattices. Early in the 'twenties some alloy specimens, made up by melting copper and zinc together were examined in this manner. The result of this investigation showed that the crystals were a random mixture of both zinc and copper atoms (figure 5·4). A year later the same samples were handed to a student for a practice run on his x-ray equipment. To everybody's surprise the result was now quite different, showing a regular pattern in which zinc and copper atoms alternated in an orderly array. In fact, the crystals had unmixed themselves, exchanging the random distribution of the two kinds of atoms for an orderly array. They had been made at a high temperature where the entropy is high and they were therefore statistically disordered. At room temperature, which is closer to absolute zero, the lower entropy means greater order and the random pattern has to give way to a regular array. However, re-arrangement of atoms in a crystal is a slow process and this is the reason why the change was only noticed a year later.

Such transformations from a disorderly into an orderly pattern

5·4 In a piece of brass the copper and zinc atoms are at high temperatures distributed at random (*left*) but take up a regular pattern when cooled (*right*).

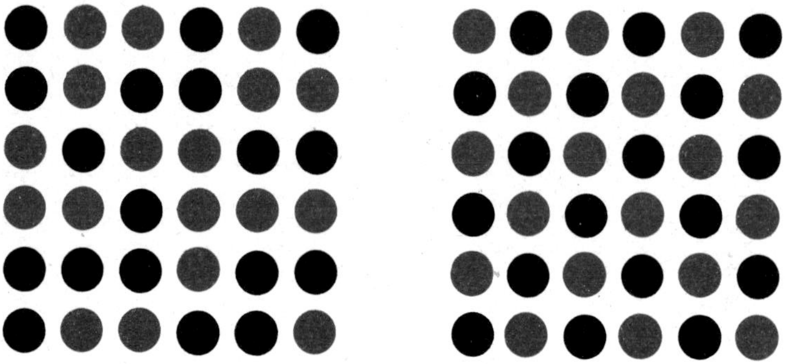

of atomic arrangement are the most striking and direct manifestation of Nernst's theorem and it is unfortunate that they are often difficult to observe because the time required for the change becomes very long at low temperatures. However, a very beautiful and impressive demonstration of the operation of the third law has been given in recent years by Peshkov and Zinovieva in Moscow. They were investigating the properties of a mixture of the helium isotopes with atomic weights four and three. The former is the natural helium from minerals or gas wells but in nuclear reactions stable helium atoms with one neutron less, and therefore the weight three can be made. This isotopic mixture remains liquid down to the lowest temperatures and, even at 1°K, it has an appreciable entropy due to the fact that the two types of atom are randomly mixed. According to the third law this entropy must disappear as absolute zero is approached. While there was still no sign of such spontaneous unmixing at 1°K, things began to change as the temperature was reduced to 0·8°K and the liquid began to divide into two phases, one richer in the light and the other richer in the heavy isotope. At 0·5 °K this unmixing has progressed so far that the two liquids are sharply separated by a meniscus between them which can clearly be seen in the photograph taken by the Russian scientists (figure 5·5). The phenomenon is as extraordinary

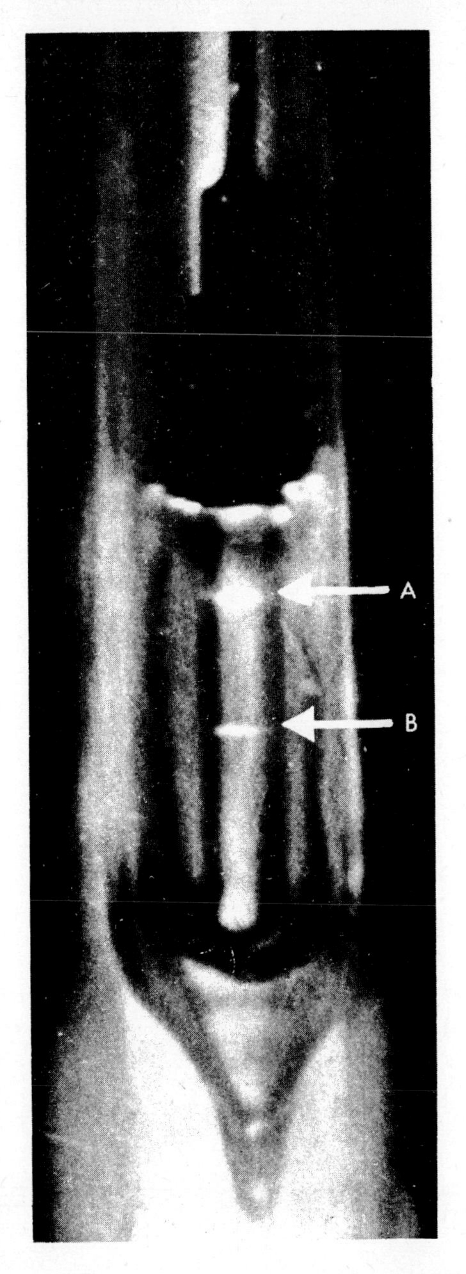

5·5 In accordance with Nernst's theorem a mixture of helium isotopes will separate into two liquid phases near absolute zero. *A* meniscus between liquid and vapour; *B* meniscus between the two liquids.

as if we were to see the red and white sand in our jar separating out before our eyes.

In our endeavour to explain the meaning of the third law we have talked about mixed crystals and liquid helium below 1 K, jumping far ahead in our story. The excuse for doing this lies in a peculiar limitation of the human brain. While we have no difficulty in recognising and defining a state of orderliness in the relative *position* of objects we cannot do the same with their *motion*. In fact, the very notions of 'order' and 'disorder' are in our mind bound up with the places which they occupy in space as, for instance, the books on a shelf or the chairs in a lecture hall. It was for this reason that we chose the ordering of atoms in the crystal lattice and the separation of the helium isotopes into two liquid phases for demonstrating the operation of the third law. However, there are other, and equally important types of ordering occurring in nature in which the regular pattern is not one of positions but one of motion.

The limitation from which our brain suffers is that we cannot form with the same ease a picture of 'before and after' as we can of 'side by side'. Somehow we have difficulties in accommodating the time on equal terms with the dimensions of space. Whether this curious limitation is inborn or a matter of education by experience is hard to say. As we shall see, it can largely be overcome by training.

We usually get around our inability of embodying the time in representations of the physical world by making a series of pictures of subsequent events. One can even achieve a sense of continuity by simplifying the progress which is being studied so much that the series of pictures become continuous. For instance, the trajectory of a shell fired from a gun can be embodied as a curve in an otherwise 'still' picture (figure 5·6). We can even go a step further by showing the simultaneous action of the target running away as soon as he sees the gun fired but the next correlation, that between the positions of shell and target at different times, cannot be easily embodied in the picture.

5·6 'Before and after' is more difficult to visualise than 'side by side'.

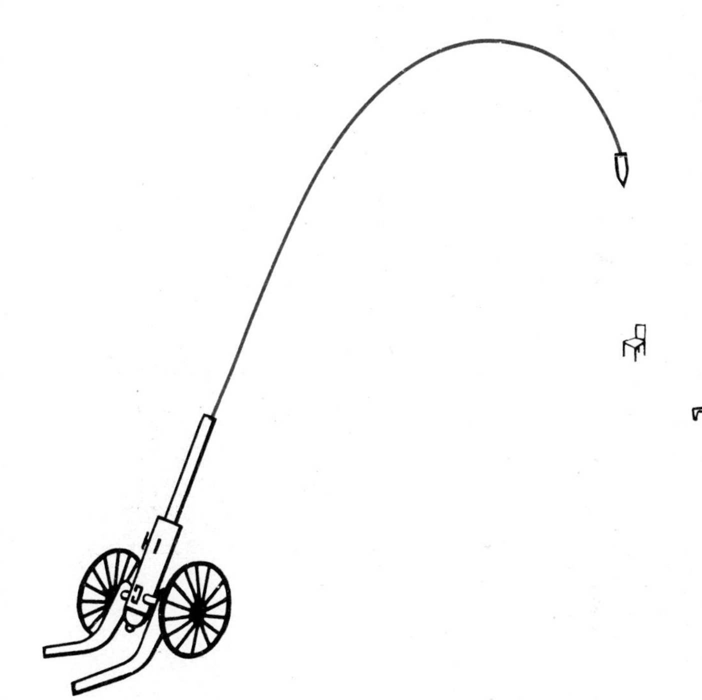

A little earlier we have, in fact, already made use of this kind of representation when the braking of the train was discussed. There we indicated the motion of the atoms by lines, each being a series of an infinite number of subsequent positions. This was further elaborated by making these lines into little arrows, a convention by which we showed that we were talking about motion and which also made a distinction between the two possible directions of motion along the line. Finally, since all our lines were meant to cover the same interval of time, the short arrows represent slow and the long ones fast motion of the atom concerned. This method of visualising the state of simultaneous motion of a number of particles can be further simplified by making the arrows all start at one point. Such a representation of events is now completely divorced from any 'side by side' pattern since the relative position of the particles has disappeared from it. Instead, we are now thinking in a three-dimensional space of velocities, a concept to which we shall return later and which we will find very helpful when trying to recognise order in the state of motion rather than in position.

Although it was eventually realised that the postulate of the entropy becoming zero at the absolute zero of temperature is the most fundamental aspect of the third law, in the beginning attention was focused on the free energy. As we have seen, this can be obtained with the aid of Nernst's theorem from the total energy of the substance. Leaving aside the strange zero point energy which will always remain in the substance, its energy can be determined in a fairly straightforward manner. The total energy contained in a piece of iron or a tumbler of water at room temperature is simply the quantity of heat which had to be put into the substance in order to warm it up from absolute zero. We have earlier come across this type of measurement when we dealt with energy in the form of a 'quantity of heat' and defined the unit, called a calorie, as that quantity which is required to raise the temperature of one gram of water by one degree.

We find that rather less heat is needed, only about one-half of a joule, to warm up one gram of iron by one degree and again another quantity, about one joule for one gram of aluminium. Since these amounts of heat are clearly specific of the substance chosen, they are called the specific heat of the material. An interesting relation between the specific heats of different substances was discovered in 1820 by the French scientists Pierre Louis Dulong and Alexis Thérèse Petit. This regularity appeared when instead of grams of weight equal numbers of atoms were considered.

The chemists of the eighteenth century had found that compounds were formed by the reaction of fixed proportions of the elements. To give an example; it was noticed that always two parts of hydrogen gas will react with one part of oxygen gas to form water, a fact which is expressed in the chemical formula for water: H_2O. If more hydrogen had been used in the reaction it would be left over as gas and the same will happen with any excess oxygen. The weights of the two volumes of hydrogen and oxygen stand in the proportion of 2 to 16 which means that two grams of hydrogen will always react with 16 grams of oxygen to form 18 grams of water. Since, as is clear from the chemical formulae, each molecule of water is formed by the combination of two hydrogen atoms with one oxygen atom, it follows that an oxygen atom must be 16 times heavier than a hydrogen atom. It also follows that one gram of hydrogen gas will contain the same number of atoms as 16 grams of oxygen gas and this gives us a means of weighing out quantities of substances, each containing equal numbers of atoms.

It is customary to start counting with the lightest atom, that of hydrogen, and we therefore have the atomic weight of hydrogen = 1 and that of oxygen = 16. By weighing the participants in chemical reactions, this determination can be extended to all elements and we obtain, for instance, figures of 56 for iron, 27 for aluminium, 64 for copper, 197 for gold and 238 for uranium.

Dulong and Petit discovered that the heat required to raise 56 grams of iron through one degree was the same as that required for 27 grams of aluminium, or 64 grams of copper or 197 grams of gold, and so on. In other words, their law says that the specific heats of elements per atomic weight, i.e. for the same number of atoms, is the same. Using the gram-atom, such as 56 for iron or 238 for uranium as unit, they found that this specific heat is about twenty-five joules per degree.

With the law of Dulong and Petit at our disposal, the calculation of energies for the third law of thermodynamics must appear a simple matter. Say we want to know the total energy of a gram-atom of iron at 0°C (=273 K) then all we have to do is to multiply the number of degrees (273) through which the iron had been warmed from absolute zero with the number of joules required for each degree (25) and we would get the result as $273 \times 25 = 6{,}825$ joules. Moreover, using a similar calculation, we should be able to give the total energy at any temperature below and above 0°C.

When in 1906 Nernst enunciated his theorem, he knew already that such simple calculations must give the wrong result. Dewar had shown that at low temperatures the specific heats were much smaller than 25 joules and that the law of Dulong and Petit ceased to hold as absolute zero was approached. In the following year the solution of the riddle was provided by Einstein. It was the first important break with the system of classical physics.

6 Quantisation

The failure of the law of Dulong and Petit became the starting point for the understanding of the peculiar position which the field of low temperatures occupies in the conceptual framework of physics. However, before one can appreciate the significance of this failure an explanation for the law itself has to be found. Again the kinetic theory provides a picture which can be easily comprehended.

The crystalline solid is, as we shall see, a somewhat complex structure of atoms which exert strong forces upon each other and it is therefore simpler to start our considerations with a gas, where these forces can be neglected. So, let us take a simple gas under simple conditions such as helium at room temperature. When talking about the van der Waals equation we have seen that the cohesive forces between the atoms become important at the critical point and below and, since room temperature is fifty times higher than the critical point of helium, we can consider the gas safely as 'ideal'. This means that we do not have to worry about cohesive effects and that it is only the energy of motion which matters. Moreover, helium presents an easy case because the gas is composed of single atoms and not of molecules, as in the more complicated cases of hydrogen or oxygen.

In terms of the kinetic picture the thermodynamic concept of temperature of the gas is explained as the average kinetic energy of the helium atoms, $\frac{1}{2}mv^2$, where m is the mass of an atom and v its mean velocity. This relation must, of course, hold for any ideal gas, whatever the mass of its atoms is, and is therefore of fundamental significance. Our unit of measurement, one degree of temperature, had been chosen quite arbitrarily as the hundredth part of the interval between melting ice and boiling water and the factor of proportionality connecting the state of motion of the gas atoms with the degree must therefore be a fundamental constant. It has been named in honour of Boltzmann and the symbol k has been adopted for it, allotting to the average kinetic energy of each of our helium atoms the value of $\frac{1}{2}kT$. In the gas the atoms are free to move in all directions of space and, since space is three-dimensional, it is convenient to express the motion of each atom in terms of the three space co-ordinates which we usually denote with x, y, and z. In other words, an atom moving through space in any direction will move a certain amount in the x-direction, another amount in the y-direction and still another amount in the z-direction. Counting up all three components of its path, its state of motion has been fully described. This makes things easy and so one has defined $\frac{1}{2}kT$ as the average kinetic energy of a helium atom at the temperature T in the x-direction. It will, of course, have an equal amount of average kinetic energy each in the y- and in the z-directions so that its total kinetic energy is $\frac{3}{2}kT$.

Using the language adopted for the motion of a particle, we say that in the gaseous state our helium atoms have three 'degrees of freedom' and that the kinetic energy for each of these is $\frac{1}{2}kT$. This notation becomes very useful when more complex cases, such as molecules made up of several atoms or matter in the solid form of aggregation are considered. Another aspect of this picture which later will acquire great significance is the so-called 'law of equipartition'. This says that the energy will be equally distributed over all degrees of freedom, a statement which must appear quite trivial in our case of the ideal gas but which has to be qualified when we begin to regard the energy as quantised.

Boltzmann's constant k refers to the energy of a single atom in the gas and it is therefore very small when expressed in the units employed in our large scale measurements which are based on human experience. Its value comes to $1{\cdot}37 \times 10^{-23}$ joules per kelvin. For purposes of measurement it is more convenient to regard not single atoms but such quantities of them which have significance on the human scale. The obvious amount of substance to be chosen is the gram-atom which we used in the interpretation of Dulong and Petit's law. By definition, the number of atoms in a gram-atom of substance is always the same, quite irrespective of whether we deal with helium, iron or uranium. It is named after the Italian Amedeo Avogadro and the symbol N is used for it. Its value is $6{\cdot}06 \times 10^{23}$. Multiplying k with N we get $(1{\cdot}37 \times 10^{-23}$

6·1 The cohesive forces hold the atoms in the crystal lattice like elastic springs.

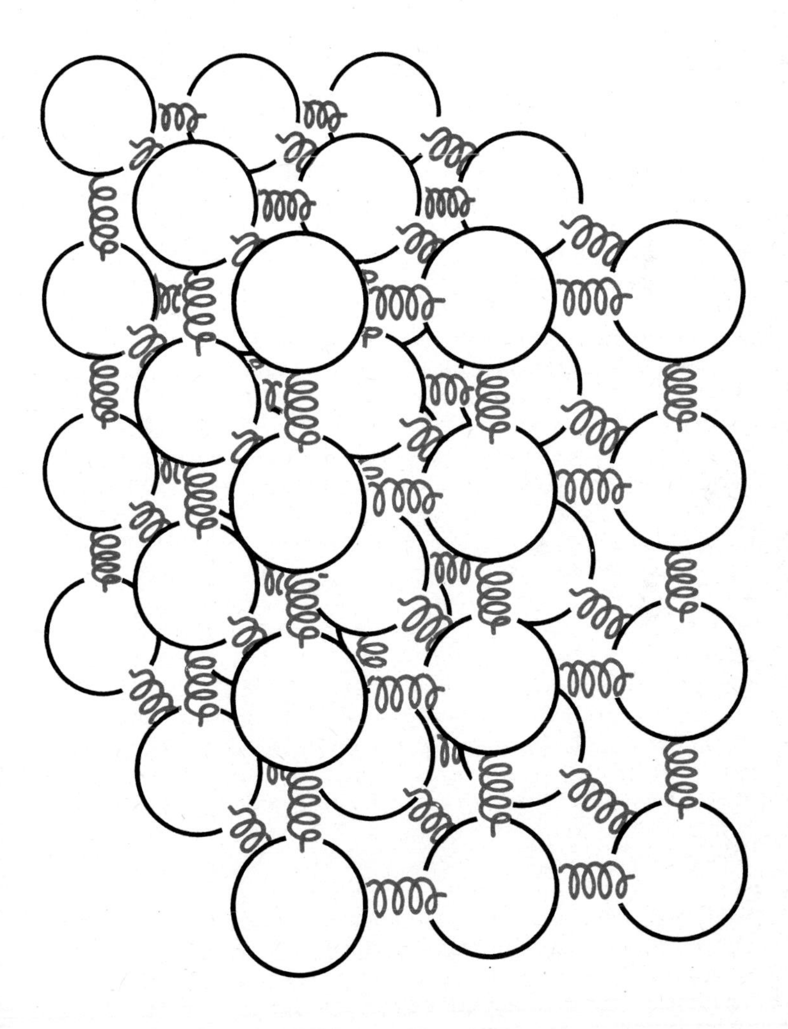

times $6{\cdot}06 \times 10^{23}$) = 8·302 joules per kelvin. This number, usually denoted by the symbol R is called the gas constant because it is the value for the constant in Gay-Lussac's law ($pv = RT$) when referring to a gram-atom of gas.

Thus the energy content of helium gas can now be given as $\frac{3}{2}RT$. This provides us straight away with the specific heat for an ideal gas, since this is the energy required to raise its temperature by one degree (change in T is 1), as $\frac{3}{2}R$. R, as we have just seen, is close to 8 which makes the specific heat of a gram-atom of helium roughly 12 joules per degree. This is, in fact, the figure obtained from experiments.

After this exercise on a simple case we can now turn to the more complicated one of the solid state. Here the atoms are arranged in the regular pattern of a crystal lattice. In it each atom is held in position by the cohesive forces exerted upon it by its neighbours. In this position the atom can move about to some extent though not as freely as in the gas or in the liquid state. In a way we can picture the atoms in the crystal as a regular array of billiard balls which are connected to each other by springs as shown in figure 6·1, representing the forces that hold the atoms in place. If in this model we pull one of the atoms a little out of its equilibrium position and then let it go, it will vibrate under the action of the springs. In the real crystal all the atoms vibrate continually and at random around their equilibrium positions and this motion represents the temperature of the solid body, just as the free motion of atoms in a gas is a measure of its temperature.

The difference between a gas and a crystal is, of course, that in the latter the atoms must return to their places. This means that in addition to their kinetic energy they also acquire potential energy. That this must be so can be seen in the simple case of a pendulum. The pendulum, too, has continually to return to its old position, swinging to and fro, held by its string and by the gravitational attraction of the earth. As it passes through the lowest point of its path, it has its highest speed, i.e. maximum kinetic energy. At each end of its swing when it is momentarily at rest it has been lifted away a little from the centre of the earth and has now acquired potential energy. In the same manner the atom passing in its vibration through the equilibrium position has kinetic energy and at the point of return it has potential energy.

Warming up a solid body therefore means increasing the degree of vibration of the atoms and we can now treat its case in a similar fashion to that of the gas. There are again three degrees of freedom for the kinetic energy because the atoms can vibrate in all directions of three dimensional space. In addition there are also three degrees of freedom in which the atoms can acquire potential energy, making altogether six. In order to heat up a solid it has therefore to be provided with an energy of $6\times\frac{1}{2}R$ per kelvin which means $(6\times\frac{1}{2}8)$ about 24 joules per kelvin. This is, of course, the magic figure discovered by Dulong and Petit and for which we have now found the explanation in terms of atomic vibrations.

We now can turn to the problem which provided the key for the understanding of the behaviour of matter near absolute zero; the breakdown of the law of Dulong and Petit. Dewar's rough measurements had shown that at low temperatures the specific heat of solids was much smaller than six calories per degree but this was not the first deviation from Dulong and Petit's law which had been discovered. As early as 1875 Wilhelm Friedrich Weber measured the specific heat of carbon, both in the form of diamond and of graphite. He found that in both cases the values obtained at room temperature were well below that of the Dulong and Petit law, the deviation being more pronounced in diamond. Extending his measurements to higher temperature he discovered that first graphite and then also diamond showed a specific heat of 25 joules per kelvin when they were heated high enough.

When a quarter of a century later Weber mentioned these results in his lectures at Zürich, they aroused the interest of a young student whose name was Albert Einstein. Einstein realised that the breakdown of Dulong and Petit's law at lower temperatures, as found by Weber in carbon and more generally later by Dewar, affected profoundly the basic tenets of physics but he could not, as yet, see what was wrong with the accepted concepts. However, at about the same time the key to the mystery was provided by Max Planck in Berlin.

Planck's paper which was destined to initiate the most profound revolution in modern physics was read before the Berlin Physical Society in December 1900. It solved one of the basic problems confronting the physicists of the day and, while Planck in the introduction left no doubt that he was fully aware of its far reaching importance, he personally disliked his own result. Max Planck was the descendant of a long and distinguished line of law givers whose spirit he had inherited. The strict and general laws of thermodynamics held for him a peculiar fascination which was based on the feeling that they had the 'absolute' character which any law of nature should, in his opinion, show. He deeply distrusted the statistical interpretation of Boltzmann's because it is based on probability and thereby seemed to him to depart from the statement of truth in its absolute form. While Planck himself never attacked Boltzmann directly, he probably encouraged Zermelo – one of his pupils – when the latter published a paper severely criticising Boltzmann's statistical methods applied to thermodynamics. It is clear that Boltzmann was right and that Zermelo, and with him Planck, had completely failed to appreciate the significance of Boltzmann's approach. In his reply, defending the statistical method, Boltzmann merely suggested that Zermelo would presumably think that the dice were loaded when he had been unable to throw a six a thousand times running, since the probability for such an event was not zero. Planck's appearance underlined his rigorous attitude of mind and the dark suit, starched shirt and black bow tie would have marked him out as a typical Prussian civil servant, had it not been for the penetrating eyes under the huge dome of his skull.

The problem to which Planck turned his mind had already defeated a number of distinguished physicists who all had attacked it because they were fully aware of its fundamental nature. Any object which is heated radiates electromagnetic energy. The radia-

6·2 Analysis of the radiation curves led to Planck's quantum theory.

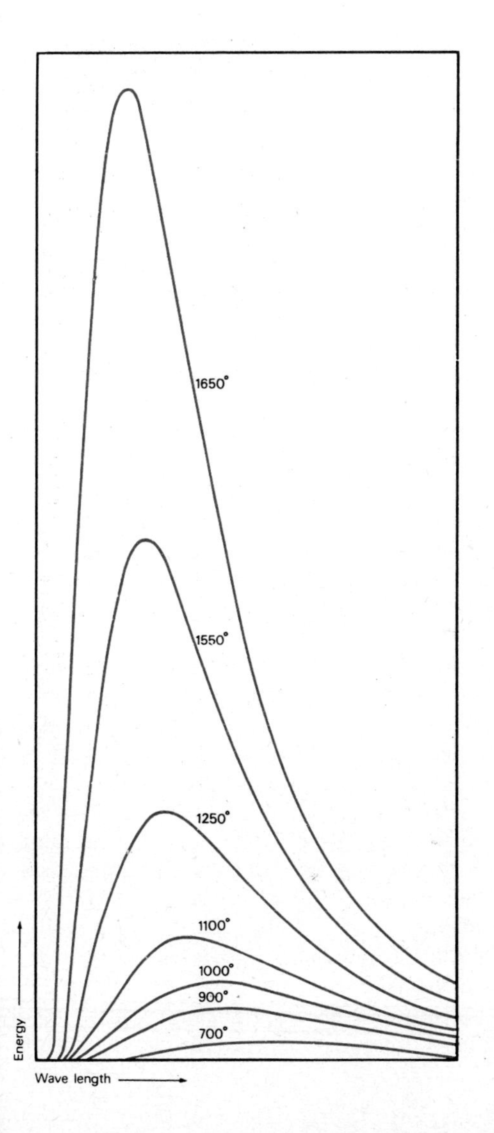

tion is given out in a spectrum of wavelengths some of which we observe as visible light or as a sensation of heat while others are registered by our instruments as x-rays, ultraviolet rays or radio waves. All these radiations are basically the same phenomenon, differing only in the length of the wave or, what comes to the same thing, in the frequency. Since all these waves are propagated with the speed of light, the frequency is simply the number of waves, of one particular wavelength, emitted per second. Accordingly, the frequency, often called the wave number, is low for a long wave and high for a short wave because fewer long waves can be accommodated in the distance travelled by light in one second than short ones.

The surprising feature of this radiation which was discovered by the experimentalists of the nineteenth century is that it does not at all depend on the nature of the radiating object. Whether it is a piece of copper or a chunk of coal or even a cloud of gas, the total amount of radiation emitted as well as the distribution of energy over the various frequencies only depends on the temperature and nothing else. This leaves no doubt that some very fundamental principle must be involved and one would also suspect that it will reveal itself in the manner in which the physical quantities involved are connected. The quantities are the absolute temperature, the energy and the frequency. Careful measurements had revealed a series of bell-shaped curves which denote the energy emitted as a function of frequency at different temperatures. All these curves have similar, though not the same, shape (figure 6·2) and the area under each curve gives us the total energy radiated at that temperature. Another feature, and one that will become of particular interest to us is that, as the temperature gets lower, not only the total energy decreases but in addition the energy is radiated more and more in the low frequencies. We are, actually, quite familiar with this phenomenon because, as we watch a hot poker cool, its colour gradually changes from orange to red, i.e. from higher to lower frequencies.

By the end of the nineteenth century a number of partial solutions of the problem had been obtained but an equation connecting all three quantities, temperature, energy and frequency, and giving solutions that fitted the experimental results was still outstanding. As more and more experimental data became available, the agreement with the existing formulae became worse instead of better and the most disturbing feature was a departure from the law of equipartition. Whereas others had based their theories on considerations of the energy Planck, who approached the problem from the viewpoint of thermodynamics, tried to relate radiation with entropy. He immediately obtained encouraging results but still the fit of his curve was not perfect and it became worse when Rubens and Hagen, the experimentalists working in Berlin, succeeded in obtaining measurements in the infra-red region of low frequencies which had hitherto been closed to investigation. It was at this stage that Planck realised the importance of Boltzmann's statistical approach which turned out to be the key to the solution. Having taken this step, he was able to write down an equation which fitted the results excellently over the whole region of temperatures and wavelength.

While Planck's solution was so perfect that there could be no doubt of its correctness, it had what appeared to him a strange blemish. The energy did not appear in it as directly related to the frequency but through a numerical factor. Planck's immediate reaction was to regard this curious relation as a mathematical oddity which could be removed by suitable manipulation of his formulae. However, what appeared to him natural turned out to be quite impossible. Whenever he had succeeded in excluding the factor from his equation, the agreement with the experimental data also disappeared. He was driven to the conclusion that this factor was essential to the interpretation of radiation and that it had fundamental significance. He had to let the equation stand in the form in which it had first emerged from his calculations. It contained the energy e in the form of the frequency, usually denoted by the Greek letter ν multiplied with a constant for which he introduced the symbol h, as $e = h\nu$.

The meaning of this relation completely changed the whole

aspect of physics. It means that the energy which until then had been regarded as a structureless fluid had an atomistic nature. As it is radiated from a hot body it does not appear as a smooth stream which can be divided up at will but one that is given out in definite units, each of which has the size of the frequency multiplied by *h*. Planck's radiation formulae implied that energy is always given out or taken up in distinct and indivisible packets which are called quanta. The size of these quanta depends on the frequency of the radiation; they are large for high frequencies, i.e. for short wavelength and small for low frequencies.

The reason why the quantised structure of energy had not been recognised earlier lies in the fact that, measured on the human scale, the constant *h* is very small. It has the form of an energy multiplied with the time and its value is of the order of 10^{-33} joules $\times$ second. This means that one calorie delivered at room temperature will consist of more than a million million million quanta. This structure of energy is so finely grained that the individual quanta will be missed by our most refined instruments for measuring quantities of heat. However, while single quanta remain unobservable and of little account in observations on the human scale, they are of paramount importance when we consider the energy received or given out by a single atom.

In our discussion of the specific heat of a solid we have interpreted the meaning of Dulong and Petit's law in terms of the vibration of single atoms within the framework of the crystal lattice. In this context the significance of the law of equipartition is that the energy can be shared out equally over all degrees of freedom in the substance. This, however, is only true under the assumption of classical physics that energy can be divided up to any desired extent. It clearly will be quite a different matter if energy is only available in the form of indivisible quanta. Here Einstein enters our story.

The quantum theory was a victory of Planck's extreme intellectual honesty over his own beliefs. The two steps which he had to take in order to arrive at his radiation formulae were equally distasteful to him. First he had to concede the correctness of Boltzmann's statistical approach and with it he had to accept the significance of probability considerations in physics which ran counter to his desire for absolute laws of nature. In the second step he had to reveal the discontinuous and atomistic nature of energy although he had always been careful to avoid even the atomistic aspect of matter since he regarded it as less basic than the continuous quantities of classical thermodynamics. The most conservative exponent of nineteenth-century physics, Planck had become a revolutionary despite himself; an innovator who had brought down the proud edifice of classical physics crashing about his ears. It is therefore understandable that while his brilliant intellect had forced him to proclaim a scientific revolution, he did little to cheer it on. In fact, nothing happened for another five years and then it was not Planck but Einstein who invoked the quantum theory.

In Einstein's mind for the first time the energy quanta seemed to take on a life of their own. He regarded them as individual 'darts of light' and used them for a beautiful and impressive explanation of the photo-electric effect. Einstein was then in Berne, and Planck in Berlin took no notice. Two years later Einstein published his brilliant explanation of the low temperature specific heats based on the quantum theory and at about the same time his first paper on relativity. Planck hailed Einstein's paper on relativity as one of the most outstanding advances in physics but remained completely silent on Einstein's success in demonstrating the fundamental importance of his own quantum theory. It has sometimes been said that Planck's disinclination to advance the quantum theory was due to his failure to appreciate its full significance. Nobody who has read Planck's first paper can believe this. The explanation for his strange attitude is rather to be found in the critical acuteness of his mind which not only fashioned the quantum theory against his own accepted views but which was painfully aware that it was incomplete. He, more than anyone else, knew that in the quantum constant must be contained a profound significance which he had failed to discover. More than a quarter of a century had to elapse until in that glorious year of 1927 the truth was revealed. In his long life, full of the most tragic personal losses, the seemingly solid world of physics and human values in which Planck had grown up collapsed but he lived to see the triumph of his idea.

We have already mentioned the problem which Einstein was the first to see; the incompatibility of the law of equipartition with energy quantisation. When a substance is heated the energy can only be taken up by the atoms in the form of fixed quanta. The peculiar nature of these quanta lies in their dependence on a frequency ν which determines their size. The temperature of the crystal lattice is given by the energy with which the atoms vibrate in it and an increase in temperature means stronger vibration. When we supply energy to a pendulum, we increase the size of its swing, its amplitude. However, as Galileo observed on a chandelier in the cathedral at Pisa, the time of each swing, the frequency, remains unchanged. Similarly the frequency of vibration of the atoms in the lattice is the same however strongly they vibrate. The frequency of vibration is the same for each atom in the crystal but varies from substance to substance. Generally speaking this frequency is higher for light atoms, particularly if the cohesive forces holding it in the lattice are strong.

Einstein made use of this model of a crystal as a large number of oscillating atoms with the same characteristic frequency and noted that it bears a close resemblance to the mathematical framework which Planck had used in the derivation of the radiation formula. There, too, deviations had occurred from the law of equipartition, especially when low frequencies, i.e. low temperatures were dominant.

Einstein based his theory of the specific heat on the assumption that the characteristic frequency of vibration, let us call it ν_0, of the atoms is the frequency which determines the size of quanta that the substance can take up. That means any atom can be made to vibrate more strongly by receiving energy impulses of magnitudes $1h\nu_0$, $2h\nu_0$, $3h\nu_0$ and so forth but it cannot accept, say,

$2{\cdot}5h\nu_0$ or, and this is of special importance, no energy smaller than $h\nu_0$.

At sufficiently high temperatures all the atoms will vibrate strongly, each with the energy of several $h\nu_0$ and there is no difficulty in sharing it out evenly over all degrees of freedom in the crystal. The situation must, however, change at lower temperatures when the average energy per atom comes down to $1h\nu_0$. From now on the energy which cannot be divided further cannot be shared out equally. It can be shown by calculation that a copper atom vibrates in its lattice with a frequency ν_0 of roughly $6{\cdot}5\times10^{12}$ swings per second. The quantities which we must compare are the energy quantum $h\nu_0$ associated with this frequency and the thermal energy per (kinetic and potential) degree of freedom kT. Since k is, like h, a universal constant, this energy depends only on the absolute temperature T. As long as it is large compared with $h\nu_0$ there is no difficulty in sharing it out equally over all degrees of freedom and the law of Dulong and Petit is obeyed. Although the energy is quantised we do not notice this because the structure of energy quantisation, given by the size of $h\nu_0$, is too fine for its grains to be noticed in the large kT.

This is no longer true, however, when the temperature is sufficiently low for kT to become of the same size as $h\nu_0$ because now a degree of freedom can receive either the minimum quantum of energy $h\nu_0$ or nothing at all. Expressing this condition by the equation $h\nu_0 = kT$, we can calculate the temperature at which it occurs as $T = h\nu_0/k$. Putting in figures for the two constants and for the frequency ν_0 in copper ($6{\cdot}6\times10^{-34}\times6{\cdot}5\times10^{12}/1{\cdot}38\times10^{-23}$ joules K^{-1}) one obtains for T the value 300. At first sight this seems to suggest that below 300 K – which is roughly room temperature – the copper crystal cannot take up any energy at all and its specific heat must suddenly become zero. However, as Boltzmann had pointed out, the nature of thermodynamic quantities is always statistical. This means that, although the size of the energy quantum remains unchanged and not all the degrees of freedom can accept one, there is always a finite probability of some of them receiving one. It is like a lottery with a motor-car to win. The car cannot be divided among the ticket holders and none except one of them will receive it but they all stand the same chance of getting it. The chance of any degree of freedom receiving an energy quantum $h\nu_0$ diminishes steadily with decreasing kT, i.e. with falling temperature. The lower the temperature of the copper crystal, the fewer quanta it can accept, which means that the specific heat decreases.

This deviation from the law of Dulong and Petit, the decrease at low temperatures of the specific heat below the value of 25 joules per gram-atom was the fact which had been observed by Weber and by Dewar. The seemingly incomprehensible phenomenon had now been explained by Einstein on the basis of energy quantisation. His theory went much further than giving a qualitative explanation. He calculated with Planck's formula the probability with which a degree of freedom would, at any temperature, receive an energy quantum and from this he could draw a curve giving the value of the specific heat as function of the temperature. As we said earlier, the solid state is a complex structure and Einstein's model was necessarily simplified. In addition to the vibration of single atoms in the lattice, groups of atoms can vibrate together and these groups may be of any size. Thus instead of only one characteristic frequency, some lower frequencies of vibration can occur in the lattice and these can accept smaller energy quanta. A somewhat better approximation was calculated seven years later by Debye but even this is still too simple to be accurate at all temperatures.

Since we are only interested in the principle of specific heat quantisation we can leave these refinements to the expert working in the field. They add nothing essentially new to the fundamental step of Einstein's theory which for our purposes is quite adequate. The formula $T = h\nu_0/k$ leads from the frequency ν_0 which is characteristic of the particular substance to a 'characteristic temperature' which has special significance for this substance. It is that temperature at which its thermal energy per degree of freedom becomes equal to the size of one energy quantum. Below this temperature the probability of degrees of freedom receiving one quantum decreases rapidly and the specific heat falls off. For copper with a frequency of $6{\cdot}5 \times 10^{12}$ swings per second, we derive a characteristic temperature of 300 K which means that below this temperature the specific heat falls off to lower values. It has already been mentioned that the frequency depends on the mass of the atom and on the strength of the cohesive forces holding it in place. The carbon atom is five times lighter than that of copper and therefore vibrates much faster. The characteristic temperature is therefore much higher. The two modifications of carbon, graphite and diamond, differ in their lattice structure and therefore in the strength of the cohesive forces. As the extreme hardness of diamond indicates, they are stronger in diamond than in graphite. This means that the frequency of lattice vibrations is higher in diamond and the characteristic temperatures of diamond and graphite are 1850 K and 1500 K respectively.

We now can understand the strange results which Weber obtained for the specific heat of these two substances. Diamond will only obey the law of Dulong and Petit at very high temperatures, close to 2000 K. At room temperature the probability of a degree of freedom receiving the high energy quantum required for it is already fairly small and the specific heat is, in fact, only a quarter of the classical Dulong and Petit value. It is, even at room temperatures strongly quantised. It is also clear why Weber found the case of diamond to be more extreme than that of graphite. On the other hand the characteristic temperatures of most other substances are below room temperature and their specific heats are at room temperature still classical, a fact which led Dulong and Petit to establish their law. An opposite extreme to diamond is lead which has a very heavy atom and in which, as is indicated by its softness, the cohesive forces are fairly small. It has a characteristic temperature of only 95 K.

While the characteristic temperatures at which the specific heat begins to drop noticeably below the Dulong and Petit value differ widely for the various substances, Einstein's theory suggests that the manner in which this drop takes place should be much the same for all of them. With Nernst's theorem focusing attention on the low temperature specific heats and with Einstein's prediction of a universal function for their decrease, the stage was set for the experimentalists. The closeness of the relevant dates is interesting. Nernst enunciated the third law in 1906, in 1907 Einstein published his theory of the specific heats, and in 1908 Kamerlingh Onnes liquefied helium.

It was Nernst's dynamic personality which led the experimental attack. He had just become professor of physical chemistry in Berlin and he lost no time in setting to work on the determination of specific heats at low temperatures. With deep theoretical understanding, bordering on intuition, he combined a great flair for elegant technique. He was an impatient man who always wanted quick results. This means that experiments which appealed to him had to be simple. He had no time for perfectionism and maintained that no effect was worth investigating if it required a better accuracy than ten per cent. He was fully aware that this was an overstatement but it well characterises his attitude. The methods used for measuring specific heats at room temperature were quite unsuitable for cryogenic work and Nernst solved the problem with the invention of the vacuum calorimeter, a simple device which in the most ingenious way made use of the physical conditions existing at low temperatures. How perfect his design was is best shown by the fact that now, more than half a century after its invention it is still used in laboratories all over the world and essentially in its original form.

Nernst travelled to Leiden to see Kamerlingh Onnes' famous laboratory. He was struck not only by its unique facilities but even more by the complexity of its equipment, and he immediately decided not to get involved in elaborate techniques of that kind. Nernst was not particularly interested in reaching the lowest temperatures but in the measurement of specific heats. It would be nice to go down to 1 K but he felt that so little was known about the behaviour of matter at, say 20 K, that liquid hydrogen would be sufficient for him. Travers in the description of this hydrogen liquefier had pointed out how cheap and simple it was. These remarks had clearly been put into the paper in order to irritate Dewar but they captivated Nernst. This was exactly what he was looking for and with his trusted chief mechanic Hoenow, he embarked on a hydrogen liquefier of his own design. Again it had a number of ingenious features and would have worked well if some attention had been paid to ancillary services such as the compressor and gas purifiers. However, Nernst could not be bothered with streamlining his equipment and instead of being a routine operation as in Leiden, hydrogen liquefaction in Berlin was an art which could only be performed with a modicum of success by the wily and experienced Hoenow. However, even twenty years later, when I had to rely for my work on their services, Hoenow and the liquefier had not got to know each other well enough to make the operation anything but hectic.

Nernst soon assembled at his laboratory an enthusiastic band of research students whom he taught physics and hard work. They cursed and admired him and proudly recalled in later years the training and inspiration he had given them. Among them were two young Englishmen: F. A. Lindemann, who, later as Viscount Cherwell became the close friend and adviser of Winston Churchill, and his brother Charles. F. A. Lindemann, in particular, took a leading part in the specific heat work and its theoretical interpretation. The brothers were excellent tennis players and devoted a certain amount of time to the game for which they were making up by working in the laboratory at night. Even so, Nernst could not quite suppress his disapproval and, showing a superb disregard for the sanctity of competitive sport, said: 'Two grown-up men chasing one little ball! You are so rich, why don't you buy one each?' Nernst himself was devoted to the country and to the shoot, and when he succeeded in selling the patent of his lamp for a considerable sum – it soon proved commercially useless – he bought an estate. His attitude to husbandry was interesting and unusual.

6·3 The drop of specific heats at low temperatures was explained by Einstein on the basis of the quantum theory.

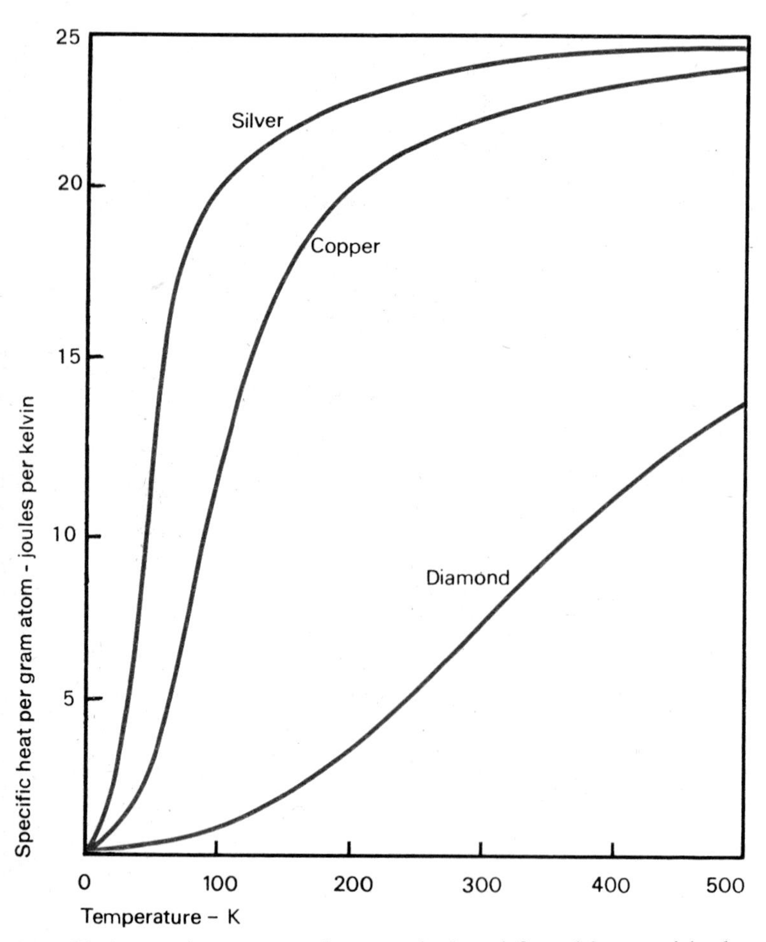

One Christmas he went to the cow-shed and found it surprisingly warm. Realising that this heat was produced by the natural metabolism of the cows, he sold the cows and instead had some ponds dug, in which he bred carp. His argument was that if meat was being produced for his money, it had to be done without a shockingly high increase in entropy.

Nernst was a man full of life, bubbling over with ideas and surprises and he left the mark of his personality on all who worked with him. The amount of work coming out of his laboratory in those eight years after 1906 was staggering. When research had to be discontinued at the outbreak of the First World War, his theorem had changed from a strange new hypothesis to a generally acknowledged and well proved fundamental law of thermodynamics. The predictions which the third law permitted had changed the aspect of chemical technology and, more important for our story, the physics of low temperatures had been given its true meaning. The path to the new concepts had been the investigation of the specific heats.

As the results began to come in from the measurements with the vacuum calorimeter, it became clear that Einstein was right. It was a triple triumph for his theory, Planck's quantum hypothesis and Nernst's theorem. As one substance after another was investigated, the general pattern emerged exactly as it had been predicted. With falling temperature, sooner or later, their specific heats dropped below the value of the Dulong and Petit law, tending gradually to vanishingly small values as absolute zero was approached. Even more satisfying, the manner in which this fall took place was the same for all substances, a gradual decline which is followed by a steep fall that eventually becomes more gradual again at the lowest temperatures. The dependence of the specific heats on temperature obeyed with astonishing accuracy the form which had been predicted by Einstein and Debye. The curves characteristic of the various substances differ widely in the temperature at which this drop takes place, depending on the characteristic frequency ν_0 with which the atom vibrates in the lattice.

A series of these curves for three of the chemical elements is shown in figure 6·3 from which it is apparent that they all have the same shape. An even better comparison between these curves can be made if, instead of plotting the specific heats of the different substances against the absolute temperature, we plot them against the temperature divided by the characteristic frequency of each substance. The plot of the same specific heats as in figure 6·3 is next shown in figure 6·4a but with (T/ν_0) instead of T as abscissa. The constant factor (k/h) is merely required to keep the dimensions correct. Now all the points for all three elements fall on exactly the same curve. They vanish as absolute zero is approached and gradually reach the 'classical' value of the Dulong and Petit law as $(kT/h\nu_0)$ becomes larger than one.

The area under the specific heat curve in figure 6·4a, i.e. the specific heat multiplied with the temperature, represents the thermal energy. The area between this curve and the 'classical' specific heat is the zero point energy. There is another way of looking at the same result but in a manner which shows more clearly the special significance which the third law gives to the low temperature region. In figure 6·4b we have plotted not the specific heat but the energy against temperature. The full curve gives the total energy of the substance. It does not start from zero at absolute zero but with a finite value which is the zero point energy. The energy rises first very slowly with increasing temperature which is, of course, due to the low value of the specific heat in this region. Then the rise becomes steeper as the specific heat increases rapidly in the neighbourhood of $(T/\nu_0)\,(k/h) = 1$ and finally the energy rises proportionally with the absolute temperature in a straight line. This is the classical range of the Dulong and Petit law where the specific heat is 25 joules per gram-atom and where each degree of freedom can accept its share of $\frac{1}{2}k$. A comparison between the related behaviour of the specific heat and the energy can easily be made because we have plotted both in the same manner against $(T/\nu_0)\,(k/h)$.

The behaviour of the energy as it would have been expected in the days of classical thermodynamics before Nernst's theorem is shown by the dotted line. It is simply a continuation of the straight line extended down to absolute zero, becoming zero at absolute zero, as expected by Amontons, and not allowing for a zero point

6·4 Energy quantisation causes deviations from the classical concepts of (a) specific heat and (b) energy, leading to a residual energy which is retained by the substance even at absolute zero.

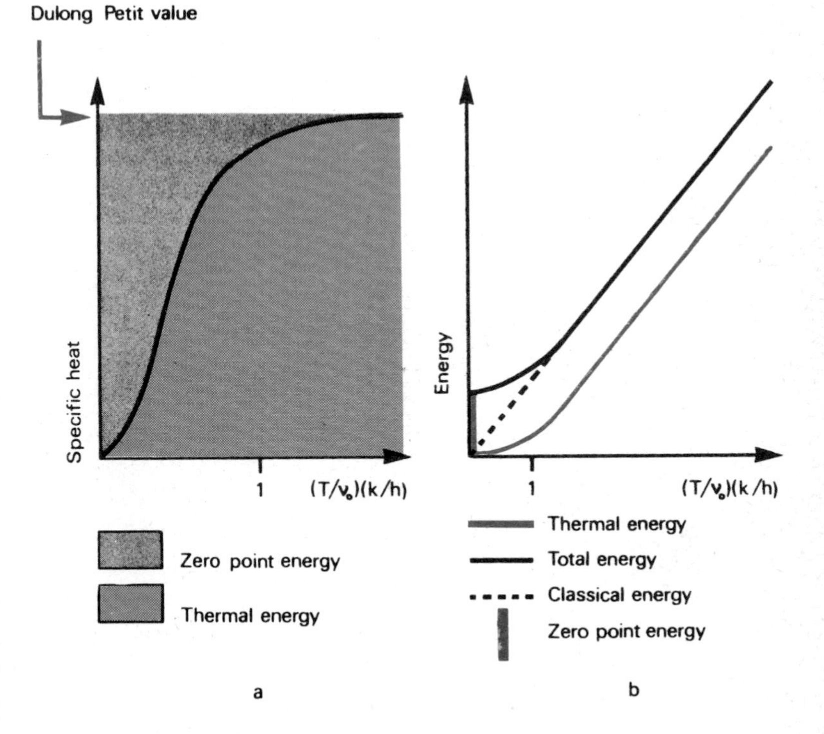

energy. The diagram also gives the thermal energy as a red curve. This is the amount of heat which has to be supplied to raise the temperature of the substance, and it is important for us because it is the quantity in our energy diagram which is actually measured. The full curve of the total energy is then constructed by adding to the thermal energy the zero point energy.

Up to now the term 'low temperatures' has been used by us in a somewhat loose way to describe a region below the range at which we live and at which we carry out most of our observations. There has been nothing to indicate that they would have any special significance, except that things are colder and, *a priori*, one has no reason to expect the aspect of matter to change basically just because it has been cooled. Here the third law of thermodynamics has made all the difference. The energy diagram shows that at high temperatures the sum of the thermal and zero point energies make up the total energy of the substance. It is clear that the existence of the zero point energy must make itself felt in the behaviour of matter because it cannot be taken out of the substance and therefore cannot take part in energy exchanges. On the other hand, at high enough temperatures its influence is not noticeable since the thermal energy has reached its classical value.

However, this state of affairs must change profoundly when we enter the region in which the thermal energy has fallen to the value of the zero point energy or even below it. It is apparent from our temperature diagram that the new, non-classical, aspect in the behaviour of matter will become serious in the neighbourhood of half the 'characteristic' temperature, i.e. near $0{\cdot}5(T/\nu_0)\,(k/h)$. Thus, with Nernst's theorem the range of low temperatures becomes well defined as the region in which the zero point energy is the dominant part in the energy content of the substance. Indeed, as Einstein showed in his ingenious interpretation of the specific heats, matter at low temperatures differs basically from its behaviour at normal temperatures because energy quantisation makes itself felt.

The way in which the quantum principle had first shown itself in Planck's radiation formula was too indirect to reveal at once the fundamental break with classical physics. Planck himself, who fully realised its implications, had done little to emphasise the new concept, partly because he disliked the break but mainly because he felt that his work was incomplete. In addition, the small value of h, giving to energy such a very finely grained structure, had suggested that quantisation would leave large scale phenomena unaffected. Nernst and Einstein had changed all this. Measurements of macroscopic size and of such conventional nature as the specific heat now turned out to be governed by quantisation. Planck was one of the first to see this and he hailed the third law of thermodynamics as the most important manifestation of the quantum principle.

Low temperature research was now established as a new subject, not only by the experimental pioneering work of Cailletet, Dewar and Kamerlingh Onnes but in the far profounder sense as a distinct subject of physics. A subject, marked out by laws of nature and by phenomena which are different from those of ordinary temperatures. Moreover, this difference is a fundamental one, the invasion of the large scale world by the rule of energy quantisation. The two salient features of the low temperature world, the vanishing entropy and the dominance of the zero point energy are its direct consequences.

7 Indeterminacy

Nernst's theorem had given a special significance to the physics of low temperature, with vanishing entropy and the zero point energy as the outstanding features. While people like Planck appreciated from the beginning the importance of the entropy concept, most physicists preferred then, as now, to think in terms of the energy. To them the existence at absolute zero of an energy which was being retained by the substance appeared as the most revolutionary departure from classical thinking. Unsuccessfully, they tried to imagine in what form this energy, which could not be counted among the thermal vibrations of the atoms, could manifest itself. For a long time its origin remained enigmatic.

Planck who had called the third law of thermodynamics the direct consequence of the quantum principle made an inspired guess which eventually turned out to be right. It was based on the way in which he had derived the radiation formula as the distribution of energy over a system of oscillators. No precise physical meaning had been given by him to these oscillators except that their function was to be the source of the electromagnetic radiations to which the formula was applied. The quantum concept then arose from the fact that the formula only worked when the energy of these oscillators was made to vary in steps which are multiples of $h\nu$. The important thing was that it had to be whole multiples and not fractions. Planck's guess was that the zero point energy of an oscillator should have the value $\frac{1}{2}h\nu$. This energy, not being a whole multiple, cannot be discarded by the oscillator and has to be retained as a lowest energy state of finite value.

Planck's guess worked well in conjuction with Einstein's theory of the specific heat and it gained even stronger ground when Bohr applied the quantum theory to his model of the atom. There the electrons are never at rest, even when the atom is in its lowest energy state: the 'ground state'. However, there is a great difference between showing that Planck's guess was right and knowing why it was right. An answer to this latter question could not be expected until some physical meaning could be given to the quantum constant itself. The success of the radiation formula and of Einstein's treatment of the specific heats, as well as a huge amount of evidence which accumulated in the following years, showed that the quantum constant must have a very profound significance in our physical world. However, for the present it was nothing more than a numerical factor which Planck had to use to make his formula fit the experimental data.

The great success of the quantum theory which was demonstrated by Einstein's and Bohr's work at first over-shadowed the emptiness of the quantum concept as such. Only Planck himself remained reticent and cautious, realising from the beginning that his great theory was sadly incomplete. As he saw it, there were two ways out of the dilemma; either the quantum concept was a mathematical oddity or it must have a deep physical meaning. For a long time he favoured the first alternative. His strong feeling that the laws of nature must be 'absolute', required relations between physical quantities which should be free from any ambiguity. Hence his preference for thermodynamic over kinetic descriptions which he distrusted because they introduced atomistic concepts. He much disliked Boltzmann's statistical approach because, to his way of thinking, it debased the simple grandeur of thermodynamic quantities by having them interpreted in terms of probability. In his search for the correct radiation formula it had been a severe blow first to realise that only the statistical approach would lead to the truth and then for himself to introduce atomistic concepts into thermodynamic quantities. Even before his first paper on the radiation formula was published, Planck had tried, though unsuccessfully, to remove the quantum concept from his work. His main reason for failing to champion his theory during the years after this first publication and for ignoring Einstein's work on it was the hope that he would find the way back to the continuity of classical physics.

As the years went by and the quantum concept scored success after success, this hope began to fade. There was now no alternative but to accept its physical reality. What had to be done was to discover its meaning. Here again, the first step was taken by

Einstein, though one wonders whether, at this early stage, even Einstein saw the opening which it provided. One year before he wrote his paper on the specific heats, he had published an application of the quantum theory to the photo-electric effect. In it he explained the liberation of electrons from the surface of the metal on to which the light was shining as due to the action of individual quanta. The concept of such 'darts of light', as he first called them, created a difficulty with which classical physics could not possibly cope. If such darts existed and Einstein's theory again provided a perfect explanation of all the observations – then they must be localised in space. Such a particle aspect of light had been suggested by Newton but had been given up long ago in favour of the wave theory. And in view of the well-known experiments on interference made by Young and Fresnel, the interpretation of light as wave motion seemed irrefutable.

The question arose of how the wave aspect of light could be reconciled with the particle nature which followed from Einstein's theory. Einstein had brilliantly explained the photo-electric effect by means of the quantum theory but at the same time he had drawn attention to a dualism of wave and particle which was to bedevil physics for more than twenty years. Were the quantised light darts, the photons as they came to be called, particles or were they waves? There seemed to be no answer to this riddle and Sir William Bragg characterised the situation well when he said the wave theory of light was being taught on Mondays, Wednesdays and Fridays and the particle theory on Tuesdays, Thursdays and Saturdays. This situation grew even worse in 1924 when a young Frenchman, Prince Louis de Broglie, suggested in his Doctoral Thesis that not only did light waves have particle aspect but that particles, such as the electron, might also have a wave nature. It is significant that this revolutionary concept came from a fresh young mind, unencumbered by ideas set in classical rigidity. All he did go by was the suspicion that nature favours symmetry and on this basis he predicted the size of the wave length of an electron travelling with a certain speed. Sure enough, these de Broglie waves were

7·1 Indeterminacy makes it impossible to locate a particle with greater certainty than that given by Planck's constant. Its de Broglie wave denotes the chance of finding the particle in a given region of space, over which it is 'smeared out' by the uncertainty relation.

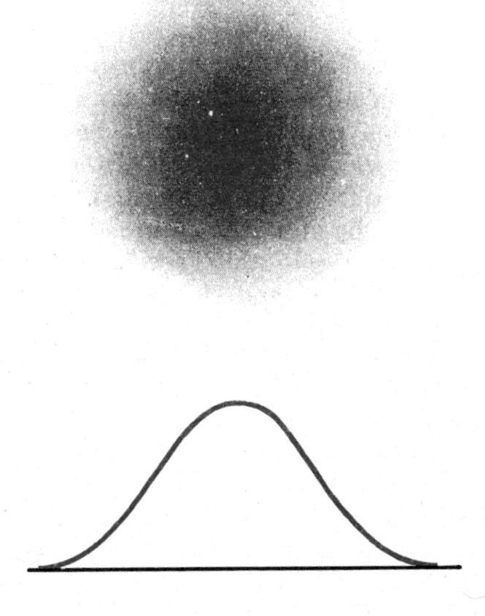

soon discovered by the experimentalists who showed that they led to interference patterns similar to those of the light waves.

For a time it seemed that physics had gone mad and that anything might be possible. The story of those years is a tale in itself which, intriguing as it is, we cannot follow here in detail. The names of those who eventually solved the problem is a list of Nobel Prize Laureates. We can do no more than mention a few highlights. In the early summer of 1925 the young German theoretician Werner Heisenberg had hayfever and went to Heligoland to clear his head. He came back with a remarkable conclusion which provided the first stepping stone; 'Only those quantities which can, in principle, be observed are allowed to be used in calculations'. This had a sobering effect on the tendency to regard, by analogy, atoms or electrons as billiard balls, only very small ones. We cannot observe the collision between two electrons in the same way as an event on the billiard table, so there cannot be any physical reality attached to the use of an analogous description.

A year later the Austrian Erwin Schroedinger transformed Bohr's model of the atom, which was a quantised solar system in miniature, into the natural diffraction pattern of the de Broglie wave of the electron around the nucleus. Planck was delighted because the wave aspect seemed to offer a return to continuity. For a fleeting moment it seemed as if the atomistic nature of both matter and energy might be resolved into smoothly varying waves. However, the clock could not be put back and classical physics did not return triumphantly carrying the quantum principle in its arms. A few months after Schroedinger's paper Max Born in Goettingen found the true nature of the waves; they denote the statistical probability of finding the particle. From that day on physicists have been content to call any small disturbance in space a particle, irrespective of whether it is an electron or a photon, the nature of the particle being described by the kind of disturbance which is created. Mass, electric charge, angular momentum and other attributes define the disturbance. All particles are associated with waves and these waves indicate the chance which we have of finding the particle in a given region of space. The crest of the wave denotes maximum probability of encountering it and if somewhere the height of the wave is effectively zero then we have little hope of meeting the particle in that place (figure 7·1).

Thus, just over ten years after Einstein had exposed the dualism of wave and particle, it was resolved by Born who showed that, when reduced to atomic dimensions, the laws of nature are always statistical. One more step remained to be taken; the quantum constant had to find its place in the framework of these new ideas. It was Heisenberg who provided the solution and it went back to the hay fever cure in Heligoland. He asked himself how closely can one locate a particle, remembering that only direct observation can have a claim to yield physical reality. Say, one wants to find an electron. Then a probe is needed to locate it. Another elementary particle, for instance a photon, can be used for this purpose but now the trouble starts. If a photon of short wavelength, i.e. with a high value of $h\nu$ is used, then its energy will alter the velocity of the electron in an unpredictable manner. Changing now to a photon of low $h\nu$, we find that its wave is so long, i.e. its own position is so uncertain, that it is useless for pinning down that of the electron with any precision.

Heisenberg could show conclusively that this kind of reasoning does not depend on the particular type of experiment which is chosen but that it is quite generally valid. He proved that, in principle, observation cannot be carried out beyond a lower limit of certainty and it turned out that this limit is given by the quantum constant h. At last the meaning of the mysterious universal constant which had first appeared in Planck's radiation formula was revealed; it denotes the limit of certainty in physics.

Heisenberg's uncertainty principle has profound consequences, not only in physics but quite generally in philosophical considerations involving the questions of determinism and free will. Any statement dealing with dimensions of h or less is merely metaphysical speculation and can never claim to have meaning as far as individual events are concerned.

The macroscopic laws of physics are fortunately saved by statistics since they always represent averages over large numbers of these individual events. However, the basis of statistical treatment was now completely changed. In the statistical approach of classical physics the macroscopic quantities are formed by averages of individual events each of which is fully determined. There statistics is merely a simplification in the use of mathematics and it is assumed that if we had adequate methods of observation and equally good means of computation, we could have dealt instead with the individual events by means of classical mechanics. In quantum mechanics the laws of physics are always the result of statistical multitudes of single events, each of which however is indeterminate. For instance, according to the uncertainty relation we are, in principle, unable to predict whether an individual radium atom will disintegrate within the next second or whether it will remain intact for a hundred thousand years. On the other hand, statistics allows us to make the precise prediction that in any lump of radium, consisting of a *large number* of atoms, half of them will have disintegrated in sixteen hundred years. In classical physics we thought that use of statistics meant that we need not bother about the individual event; quantum physics has taught us that we cannot know about it.

For our purposes the uncertainty principle is important because it leads directly to the mysterious zero point energy. In fact, its enigmatic character is now quite lost since it appears as the direct consequence of indeterminacy. Heisenberg showed that, when trying to observe a particle, we will either make its position uncertain when we measure its velocity or, conversely, make its velocity uncertain when determining its position. The uncertainty thus arises as a product of these two quantities. Actually, instead of dealing with the velocity v, we must use the momentum mv, thereby making allowance for the mass m of the particle. The uncertainty relation then reads:

$$\Delta mv \times \Delta l \gtrsim h$$

where l has the dimension of a length, i.e. our yardstick for measuring position. The Δ signs are used to indicate that we are dealing with *changes* in momentum or position. The equation thus says that we can only observe such changes provided their product is larger or of the same order as the quantum constant h. The limit of observation is therefore set by $mv \times l \sim h$ or, what comes to the same, $mv \sim h/l$. Squaring this and dividing the equation by m, we get $mv^2 \sim h^2/ml^2$. Now the left hand side of our equation is a mass multiplied with the square of a velocity and this quantity, as we have seen in earlier chapters, denotes the kinetic energy of the particle with mass m. It shows that the uncertainty in locating the particle within the distance l must, in some way, confer the energy h^2/ml^2 upon it.

7·2 Indeterminacy confers an energy on a particle confined into a box.

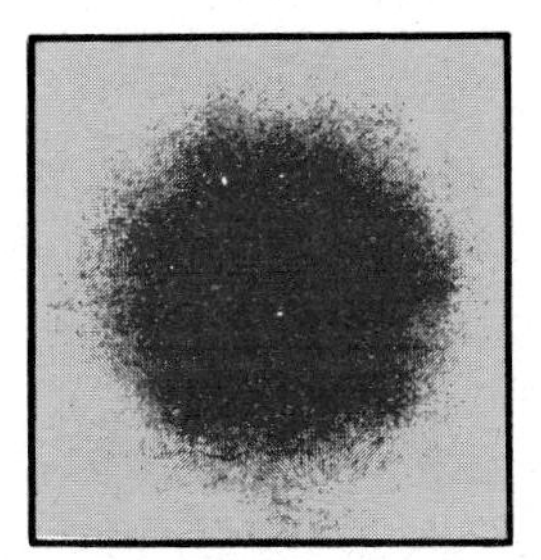

A simple way of looking at the problem is by seeing what happens when we confine a single particle into a space of diameter l. We now know that the particle is somewhere inside this box but the uncertainty principle does not allow us to find out exactly where it is at a given time (figure 7·2). Anywhere in the box there is a finite probability of encountering the particle and this probability is given by the de Broglie wave, which for instance, tells us that we have a better chance of meeting it in the centre than at a given place near the wall. This comes to the same as saying that, in a manner which is undetermined in detail, the particle vibrates inside the box, visiting any part of it at some time. Moreover, it can be seen from our equation that, if l is made smaller, i.e. if the particle is confined into a smaller box, its energy will become larger and it must vibrate more violently, the same thing will happen if m is made smaller, i.e. if a lighter particle is placed in the box.

The vibration of the particle by virtue of the fact that it may occupy any part of the box is therefore a direct consequence of indeterminacy. This energy which the particle *must* have, simply because a certain space is allowed for its motion, is its zero point energy. In a crystal each atom is confined to a 'box' formed by its neighbours from which it cannot escape and which it alone occupies. The atom has therefore, even in its 'ground state', the energy h^2/ml^2, where l is now the distance between its neighbours. Also, since this energy arises solely from the space allowed to the atom within the crystal lattice, it cannot be given up.

The uncertainty principle not only explains the existence of the zero point energy, it also accounts for its nature. We have seen that the quantum statistical probability of finding the atom anywhere in the space allowed to it by its neighbours is equivalent to its vibration within this space. The zero point energy is, therefore, in kind, indistinguishable from the thermal energy which also is represented by atomic vibrations.

Direct evidence for the idea that the zero point energy of a crystal shows itself as vibration of the atoms was provided in 1927, the same year in which Heisenberg had enunciated the uncertainty principle. It came from an x-ray investigation on rock salt, carried out by R.W.James and Miss E.M.Firth at Manchester. The discovery that x-rays falling on to a crystal are diffracted by the regular pattern of the atomic lattice had been made in 1912, independently by Max von Laue and by the Braggs, father and son. The new method of research opened the door to the investigation of the microscopic structure of matter in the realm of atomic dimensions. Earlier we came across its use in the observation of ordering which takes place in mixed crystals. X-ray diffraction gives information on the position of atoms in the crystal with respect to each other and, for instance, reveals the difference in structure between diamond and graphite which are two different types of lattice, both composed of carbon atoms. However, the atoms are not at rest. They vibrate around their equilibrium positions under the influence of their thermal energy. This thermal motion always makes the x-ray patterns which are obtained, to some extent, fuzzy. Since at low temperature the thermal motion is decreased, the fuzziness will be lessened. The Manchester experiment was therefore designed to investigate the rock salt crystal when it was at liquid air temperature. In the following year a careful mathematical analysis of the results was published in collaboration with Waller and Hartree which showed unambiguously that the atomic vibration recorded at low temperatures was much larger than could be accounted for by the thermal energy of the crystal. The zero point energy, showing itself as atomic vibrations, had been recorded in a direct experimental observation.

Even well before this time people began to suspect that the zero point energy might manifest itself in the behaviour of matter at low temperatures. Already in 1916 Nernst had drawn attention to the strong deviations from Trouton's rule which had been found in liquid hydrogen and even more so in liquid helium. This rule which had been used by the pioneers of low temperature research to forecast the behaviour of the gases which they planned to liquefy concerns the heat of evaporation, i.e. the energy which is required to turn the liquid into vapour. This extra energy is needed to impart to a molecule in the liquid a velocity high enough for letting it break away from the cohesive forces of its neighbours. Trouton had shown that this energy is proportional to the absolute temperature, i.e. the lower the boiling point of a substance, the smaller will be the heat of evaporation. That is, of course, quite reasonable when we remember that a substance has a low boiling point simply because the cohesive forces between the molecules are small.

Dewar foresaw this difficulty when with special care he shielded against heat influx the cryostat in which he was going to liquefy hydrogen. On the basis of Trouton's rule he could expect hydrogen to have a four times smaller heat of evaporation than oxygen which means that it would evaporate very easily. What he did not expect and what he could not explain was that the heat of evaporation was twice smaller than forecast. An even worse departure from Trouton's rule was found by Kamerlingh Onnes in helium.

Nernst's suggestion to look upon these deviations as due to the zero point energy was taken up by his pupils Bennewitz and Simon seven years later when more experimental data on the condensed gases had become available. They used these results in a calculation by which they showed that the unexpectedly low values of the heats of evaporation can indeed be explained through the influence of the zero point energy.

Trouton's rule is valid at normal and high temperatures and is based on the fact that the total energy of the substance is roughly equal to its thermal energy. This is the energy available for evaporating the liquid and it is roughly proportional to the absolute temperature. Things change, however, as absolute zero is approached because now an appreciable part of the total energy is made up of the zero point energy which remains in the substance and does not decrease with falling temperature (figure 6·4b). As a consequence of this, there is comparatively more vibrational energy available than according to classical physics and this extra energy facilitates evaporation. This means that the heat which has to be supplied for evaporating the liquid is correspondingly smaller than had been expected on the basis of the classical calculation.

The same as is true for the heat of evaporation also holds for the energy required to melt the substance. An extreme case is reached in helium. Kamerlingh Onnes could not understand that even at the lowest temperature which he attained, only 0·8 degrees above absolute zero, helium refused to solidify but remained a liquid. We now know that even at absolute zero helium will be a liquid and that an external pressure of 25 atmospheres is required to transform it into the solid state (figure 10·3). This state of affairs is quite inexplicable in classical physics where one would expect any substance on cooling first to become liquid and ultimately to freeze into a solid. Simon showed that the strange behaviour of helium is a direct consequence of its high zero point energy. This energy is big enough to outweigh the small cohesive forces of this substance so that the zero point vibration of the helium atoms, even at absolute zero, keeps them too far apart to form a crystal. Only the application of an extra pressure from outside will bring them near enough together to give them a chance of linking up into a solid lattice. Even well above absolute zero the effect of the zero point energy on the nature of liquid helium is felt in its anomalously low density, another feature which Kamerlingh Onnes noted on the day of first liquefaction.

There exists another case of even higher zero point energy than in helium but in order to appreciate its unique significance we have to go back to the early days of the formulation of the third law. The particular way in which Nernst had derived his theorem, by considering chemical equilibria, had left it with an unfortunate blemish. He could only postulate its validity for matter in the condensed state, that means for liquids or solids but not for gases. The trouble was that the specific heat of an ideal gas must, according to classical physics, remain finite down to absolute zero. It can be argued, and it was, in fact, argued that as absolute zero is approached all substances will pass into the condensed state and that for this reason the concept of an ideal gas had little meaning. But Nernst would have none of that; he considered this kind of reasoning a subterfuge, unworthy of his theorem. He maintained that the latter is a fundamental law of thermodynamics and as such cannot be subject to any limitations. The line of argument which he pursued was typical of the man and of his unbounded confidence in his hypothesis. Since, he reasoned, his theorem must be universally valid, the classical law of specific heat for an ideal gas must be limited. He therefore predicted that the specific heat would decrease as absolute zero is approached and this would be achieved by an as yet unknown process which he called 'gas degeneracy'.

Many of Nernst's colleagues and even those who admired him were sceptical about this abstruse phenomenon which sounded more like a prophecy than a scientific prediction. Nernst remained quite unmoved and he had Planck on his side. Planck was a notoriously cautious man who, when asked a question by one of his research students, would invariably reply: 'I will give you my answer tomorrow'. However, he knew no caution where the laws of thermodynamics were concerned, since he considered them as certainty. Planck was certain that gas degeneracy must exist and be concomitant with the quantum principle but he could not see any way of putting it into his theory. After all, the energy of an ideal gas lies entirely in the straight motion of its atoms and it appeared impossible to assign to it a frequency of oscillation which can be multiplied with h in order to make a quantum. The 'Sitzungsberichte' of the Prussian Academy of Sciences are punctuated with desperate attempts by Planck and others at finding a solution.

We have seen earlier that Boyle's law for an ideal gas can be derived by applying statistical methods to the motion of the gas molecules. The method used in the nineteenth century by Maxwell and Boltzmann is, of course, based on classical mechanics and it was therefore not too surprising that it failed to give any indication of gas degeneracy which has its origin in energy quantisation. The way in which the problem of gas statistics is attacked mathematically consists in dividing up the volume occupied by the gas into small compartments or 'cells'. Then the probability of finding the particle in a particular cell is evaluated. The same is done in a space made up of the three dimensions of velocity and, for reasons of mathematical convenience, the computations are in one operation carried out in six-dimensional 'phase space', embracing three position and three momentum co-ordinates. Incidentally, there is nothing mysterious about a 'space' of more than three dimensions which is simply introduced for greater ease of simultaneous mathematical treatment of several quantities describing a particle. Six-dimensional phase space was quietly introduced at the end of the last century by Willard Gibbs without philosophers or occultists taking any notice. The size of these cells in classical phase space is arbitrary but it is clear from Heisenberg's uncertainty principle that arbitrary subdivision is not possible in quantum physics. Since the limit for determining an individual event is set by the quantum constant h, physical reality cannot be claimed for anything happening below it and the size of a cell in quantum statistics therefore becomes h^3.

Thus, in the quantum statistics of an ideal gas we have now separate energy states, just as Einstein, in his theory of the specific heat of a crystal, had assigned quantised energy states to the vibration of the atoms. This means that the specific heat of an ideal gas must be quantised in a similar manner and become zero at absolute zero. With this step Nernst's prediction of gas degeneracy had been given firm theoretical justification based on the quantum principle.

We have explained the quantisation of molecular motion as a

necessary consequence of the uncertainty relation because that is the most straightforward and simple form of reasoning. It should, however, be mentioned that historically quantum statistics preceded indeterminacy by three years. In the summer of 1924 Einstein received for translation a short paper written by a young Indian physicist in Dacca, S. N. Bose. Bose felt that Planck's classical derivation of the radiation formula should be replaced by one based solely on statistics and the quantum principle. He had been able to do this by introducing phase cells of dimension h^3. In a postscript Einstein drew attention to the great importance of Bose's paper since it opened the road to a quantum statistics of material particles. He then developed the idea himself in two contributions to the 'Sitzungsberichte' in what became known as Bose–Einstein statistics, as distinct from the classical Maxwell–Boltzmann statistics.

A certain basis for the new statistical treatment had already been given by Planck in his treatment of oscillators which contained an important departure from classical statistics when it was later applied to molecules. In Boltzmann's method of counting, he differentiates between individual molecules whereas, in fact, they are indistinguishable. For instance, if we have two cells A and B, and we want to place two particles a and b into them, classical statistics yields four possible combinations, namely:

	A	B
1.	a	b
2.	b	a
3.	ab	–
4.	–	ab

Now, if a and b are two hydrogen molecules, they are indistinguishable and there is no sense in differentiating between them. Calling them both x, Bose–Einstein statistics only permits the three combinations:

	A	B
1.	x	x
2.	xx	–
3.	–	xx

Normally, in classical physics we do not notice the difference in counting between the two types of statistics because many more energy states are available than there are particles to place into them. In this way one never notices the excessive number of combinations given by classical statistics because so many cells remain empty anyway. Things change, however, at low temperature where the thermal energy is low and most of the cells tend to be occupied.

By 1924 gas degeneracy had been transferred from the status of a vague hypothesis to a sound theoretical basis but experimental verification was as far off as ever. In fact, it had become even less likely. Quantum statistics had, at the same time, established gas degeneracy as a real effect and shown that it would almost certainly be unobservable in any real gas. According to the theoretical prediction, the deviation from the ideal gas law must depend on a 'degeneracy parameter' which has the form $Nh^3/V(2\pi mkT)^{3/2}$ where N is Avogadro's number, V the volume occupied by one gram-atom and T the absolute temperature. In order to make this parameter appreciable, V, m and T must be small, i.e. one should observe degeneracy best in a dense, light gas at low temperatures. The obvious choice is helium and the largest effect can be expected at the critical point because at lower temperatures the gas cannot be made very dense without liquefying. Even under optimal conditions the deviation from the ideal gas law, due to degeneracy, is only about one per cent. This is much smaller than the influence of the cohesive forces between the helium atoms which is difficult to assess accurately and by which degeneracy will be completely swamped.

For a while it looked as if Nernst's prediction of gas degeneracy, though true, might never be verified. Then it was suddenly realised that gas degeneracy is responsible for one of the best known phenomena in physics: the electrical behaviour of metals.

Through the work of Faraday, Maxwell and many others electricity and magnetism had, in the nineteenth century, become a well established subject in the general fabric of classical physics. At first the similarity in behaviour between a current of water and one of electricity had suggested that charge might be a homogeneous fluid. However when in 1897 J.J. Thomson discovered the electron, electricity was recognised as atomistic in nature and the electron accepted as the carrier of charge. From there it was only one more step to suggest that the difference between an insulator and a metal lies in the fact that in the latter electrons are free to move about and able to form a current of electricity.

This step was taken at the turn of the century by Drude who formulated the concept of a 'gas' of freely mobile electrons in a metal, a theory which was further developed by the great Dutch theoretician Hendrik Antoon Lorentz. Strong support for the existence of the electron gas was provided by the Rutherford–Bohr atomic model which described the atom as consisting of a positively charged nucleus which is surrounded by negative electrons. The theory assumed that, for reasons which became apparent only much later, in some solids one or more electrons per atom become detached from the confines of atomic structure and are free to move throughout the volume of the substance like the molecules of a gas inside a container. If such a substance is placed between the positive and negative terminals of an electric battery the electrons will be attracted by the former and repelled by the latter, producing an electric current. Equally, an electromagnetic wave of light impinging on the substance will cause a corresponding movement of free electrons in the surface which does not allow the wave to enter but, instead, reflects it back. In other words, a substance to which we ascribe a gas of free electrons shows exactly the behaviour which we associate with a metal.

As can be seen from all this, the electron theory of metals fits so well that one suspects it must be right. It was therefore disconcerting when two rather fundamental discrepancies were discovered. First of all it appeared from the magnitude of the metallic heat conduction that the electrons must all have a very high average velocity which does not depend on the temperature, indicating that here the gas laws were not applicable. The second discrepancy concerned the specific heat. We have seen earlier that the three kinetic and three potential degrees of freedom for the atomic vibrations yield a specific heat of

$$\frac{6}{2}R$$

per gram-atom which is equivalent to about 25 joules. If now in the case of a metal there are in addition free electrons, say one per each atom, moving about like molecules in a gas, they should contribute another three kinetic degrees of freedom, making the total specific heat

$$\frac{6}{2}R + \frac{3}{2}R = \frac{9}{2}R$$

i.e. about 37 joules. However this, of course, is not true and the same specific heat of 25 joules is found for metals and insulators.

The solution of this riddle was eventually provided by Nernst's gas degeneracy and it is an interesting example of how the final explanation is often arrived at by the efforts of a number of scientists, each making a significant contribution. In 1925 Wolfgang Pauli postulated the important principle of quantum mechanics which bears his name. He interpreted the spectroscopic data in terms of a new law of nature according to which in any one atom no more than two electrons can have orbits of the same quantum energy. Even these two must be spinning in opposite directions, i.e. they must have angular momenta of opposite sense. We cannot here enter into a discussion of the profound meaning which this 'exclusion principle' has in the framework of modern physics. It shows that 'spin' in the world of elementary particles has a far greater significance than the angular momentum of classical physics. It is a basic property of the particle such as its mass or electric charge. There is further the curious fact that the spins somehow *know* of each other in a way which forces them to avoid the same quantum state. By using relativistic considerations, Dirac has embodied this mutual knowledge into a consistent pattern of quantum mechanics but even so it remains to some degree enigmatic.

In the following year Enrico Fermi in Italy went a step further and extended the Pauli principle from the confines of a single atom to a much wider physical system, such as a gas. In the opening paragraph of his paper he refers to Nernst's gas degeneracy, making the exclusion principle the basis of a new statistical treatment. Independently, the same step was taken a few months later by Dirac. The essence of Fermi–Dirac statistics is the mutual knowledge which the gas molecules have of each other and which, according to the Pauli principle prevents two molecules in a volume of gas from having the same energy state. This means that the new manner of counting up statistical probabilities is not only different from the Maxwell–Boltzmann method but also from that of Bose and Einstein because now only one particle can occupy each cell in quantised phase space. Using the same example as on page 150 there is in Fermi–Dirac statistics only one possibility in which two particles can be distributed over the two cells A and B which, of course, is:

A	B
x	x

This leads at low temperatures to a quite different situation from those postulated by either classical or Bose–Einstein statistics. Since there is only room for one particle in each cell an enormous number of cells will be filled up even at absolute zero and this fact alone leads to a high zero point energy. From his model Fermi also calculated the specific heat of a highly degenerate gas and found that it should be proportional to the absolute temperature at low temperatures. Eventually, it will level off to reach the constant classical value of $\frac{3}{2}R$.

If Fermi had gone a step further, he also went a step too far, suggesting that Bose-Einstein statistics was only applicable to photons and that all material particles should obey the new statistics. He clearly did not realise at this stage that in the Pauli exclusion principle *the spin* was the significant feature which governs the distribution of electrons over the orbits around the atomic nucleus. The gas to which he applied his calculations was helium. It has since become clear that only particles with odd numbers of spins will obey Fermi–Dirac statistics while those with even numbers can always have two opposite spins cancelling each other's angular momentum. Helium has two orbiting electrons and four particles in the nucleus, two protons and two neutrons, which all have spin. Thus, the number of spins in helium is even and its atoms obey Bose–Einstein statistics. This fact is mentioned here because it will become of great importance in the discussion of the peculiar properties of superconductors and liquid helium.

After another year there was a new paper by Pauli who refers to Fermi's work and now applies the new statistics not only to the electrons orbiting around an atom but to the electron gas in a metal. Strangely enough he did not take the obvious step of applying Fermi's formula for the specific heat to it. Instead, he used the new statistics for an explanation of the weak diamagnetism of metals which had been a riddle for a long time and which he now was able to explain. Perhaps his failure to calculate the electron specific heat was due to the extreme caution with which he approached the problem of the electron gas. In the introduction to this paper he is at great pains to point out the tentative nature of his suggestion.

Another year passed, it was now 1928, until the final step which solved the whole problem of the electron gas was taken by Arnold Sommerfeld in Munich. Sommerfeld combined the exclusion principle, Fermi–Dirac statistics and Pauli's suggestion to apply it to the metal electrons. The result was the famous paper which is quoted in every textbook on the subject and which at one stroke solved a number of outstanding questions.

First of all we must go back to the degeneracy parameter mentioned on page 150. Applying it to the electron gas, we find that it is extremely large because the mass, m, of the electron is so small; it is seven thousand times lighter than the helium atom. In addition, the electron gas is very dense since the electrons are held together by the positive nuclei of the atoms in the narrow confines of the crystal lattice, and this makes the volume V occupied by a gram-atom of electrons very small. We see that both m and V are in the denominator of the degeneracy parameter which thus becomes enormous. Secondly, the electrons obey Fermi–Dirac statistics which, as we have seen, must give the electron gas a high zero point energy.

Evaluating all these factors, Sommerfeld could show that the electron gas is not only degenerate but that its specific heat will not reach the classical value of $\frac{3}{2}R$ until a temperature of about 30,000 degrees is attained. Since all metals evaporate long before such a temperature is reached, the electron gas never even approaches its classical value. At normal temperatures it is already completely degenerate and its specific heat is too small to be detected. In this way Sommerfeld's paper not only explained the missing factor of $\frac{3}{2}R$ in the specific heat of metals but he also showed that gas degeneracy, as postulated by Nernst, exists and even in an extreme form. All that now remained was some direct experimental proof of the theory. It was provided through the means of low temperature research.

The trouble with observing the electronic specific is that, already at normal temperatures, it is so very small. Since, according to Fermi and also to Sommerfeld, it should be proportional to the absolute temperature, it will, at 3 K, be still a hundred times smaller. However, not only the specific heat of the electrons diminishes with falling temperature but that of the crystal lattice, too. Fortunately, it turns out that at the lowest temperatures the specific heat of the lattice vanishes more rapidly than that of the degenerate electron gas. This means that, provided the measurement is extended to a low enough temperature, the electron specific heat must become noticeable, and eventually even dominant. This state of affairs is reached at the temperatures of liquid helium and the predicted effect has since been found in all metals (figure 7·3). In this region the specific heat of the lattice is proportional to the cube of the temperature while that of the electrons remains proportional to it. For those who like mathematical equations we can express the specific heat of the metal as $AT^3 + BT$, where A and B are constants and the first term accounts for the lattice while the second is due to the electrons. Whatever size the constants may have, at low enough values of the temperature T the second must become larger than the first.

The most convenient way of visualising the properties of the degenerate electron gas is by looking at the electrons not in the conventional space of positions but in velocity space. This method of representation has been mentioned earlier, in chapter 5, when we discussed the degree of order in a system. Plotting the positions of the individual electrons in the metal would give us little useful information; they are randomly distributed within the space available to them. Instead, in velocity space, we plot their velocities, i.e. the distance covered by each electron in unit time, all starting at one point, as shown in figure 7·4. In a gas of perfectly free electrons, those in a particular velocity range will move with equal probability in any direction. Having equal velocity, they will, in unit time, all travel equal distances from the common origin. Their positions in velocity space must therefore lie on the surface of a sphere with the origin at its centre.

At absolute zero, the electrons obeying Fermi–Dirac statistics must fill all the lowest available energy states, each being occupied by two electrons with opposite spin. Accordingly, there will be a sharp cut-off at a given limiting velocity. We will find electrons in all permitted quantum states of lower velocities and none with higher ones, and in our representation the limiting velocity now corresponds to a sphere with a sharply defined surface. This is called the Fermi surface and the size of the sphere is a measure of the zero point energy of the electron gas.

As the temperature of the metal rises above absolute zero, some

7·3 Near enough to absolute zero the specific heat of the electrons in a metal will always outweigh that of the crystal lattice.

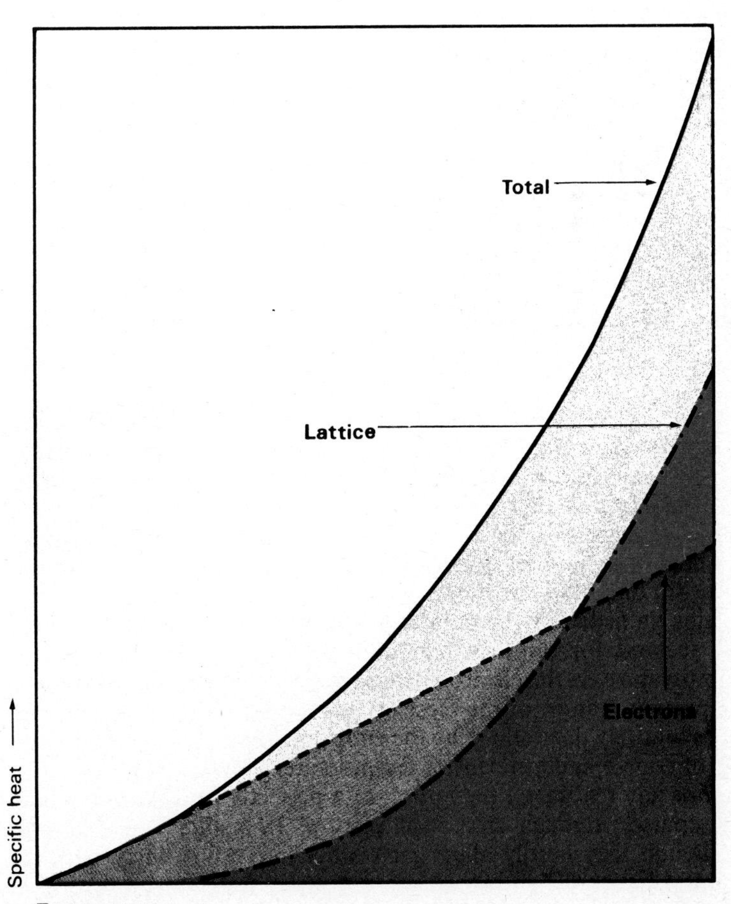

7·4 In the space of velocities, the metal electrons occupy a sharply defined sphere at absolute zero (*left*). As the temperature is raised, the surface of the sphere becomes fuzzy (*right*).

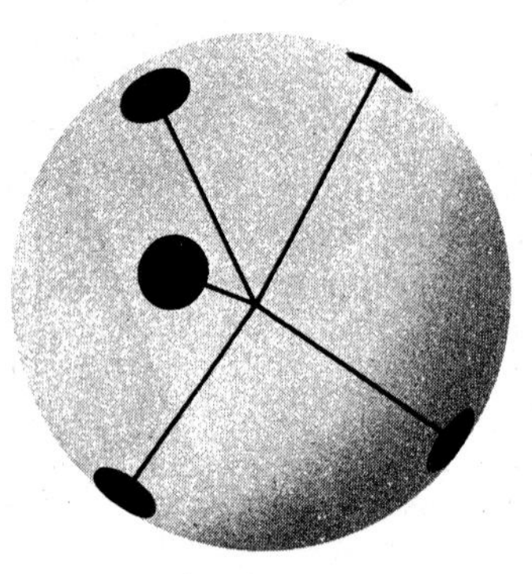

of the electrons will acquire higher velocities than the limiting one, and so some of the lower energy states inside the sphere will become empty. This means that the Fermi surface, instead of being sharply defined becomes a bit fuzzy. However, since the gas is highly degenerate, the increases in velocity caused by the supply of heat is very small compared with the high zero point velocity. The fuzziness of the Fermi surface will be only slight. This explains immediately the failure of Drude's classical electron theory to account for the heat conduction of metals; the influence of temperature on the electron velocities is negligibly small. The most important feature of the electron gas is that its properties are almost entirely determined by the enormous zero point energy which is, of course, independent of temperature.

For any real metal the notion of a perfectly free electron gas is, of course, a much simplified picture. In reality the electrons, although they hardly affect each other interact strongly with the crystal lattice. They cannot move equally freely in all directions and, far from being a sphere, the Fermi surface is usually heavily distorted. Exploration of the Fermi surfaces of various metals has occupied the main efforts of Shoenberg in Cambridge and of other low temperature laboratories in the course of the last decade. Often the real Fermi surfaces found by them look like pieces of modern sculpture or, as has been said jokingly, like monsters from velocity space. It is interesting to note that for the monovalent metals, sodium and gold, the simple theory is obeyed remarkably well and their Fermi surfaces have turned out to be close to a spherical shape. Low temperatures offer the only possibility of this type of study because close to absolute zero the surfaces are hardly fuzzy and can be investigated in detail. However, we cannot here deal with the results obtained, the appreciation of which would require a far reaching excursion into the theory of metals. In any case they are not particularly significant for low temperature physics since the electron gas is still heavily degenerate even at room temperature.

Returning to the question of the specific heats, that of the metal electrons shows the approach to zero entropy at ordinary temperatures which for them, because of their very high zero point energy, are already 'low'. Since we have defined 'low' temperatures as those at which the zero point energy becomes dominant, a certain temperature which is low for one physical system will be high for another. 'High' is any temperature for a system if something significant can still happen to it at a lower one. In fact, although we know that its entropy must become zero at absolute zero, we can never be quite sure that at a temperature which is considered by us as low the entropy of the system is already low enough to allow extrapolation to absolute zero. Sometimes there are indeed indications that some important change in a substance lies hidden at a still lower temperature.

In the early 'twenties the specific heat of solid hydrogen had been measured in Nernst's laboratory in Berlin down to 10 K and had been smoothly extrapolated to absolute zero. However, the thermodynamic data for hydrogen derived from this work did not agree with those obtained from chemical reactions. The suspicion arose that something important must still occur in hydrogen between 10 K and absolute zero. It was the object of my doctoral thesis to solve this riddle and so the specific heat measurements had to be extended to lower temperatures. The results fully justified the suspicion; at 6 K it had become clear that the specific heat did not drop as rapidly as had been expected and at 3 K it even began to rise with falling temperature. At that time it was impossible to follow the effect to temperatures lower than 2 K but since then improved technique has revealed a large peak in the specific heat of solid hydrogen near 1 K. The cause of this anomaly is closely linked with the high zero point energy of solid hydrogen which shows itself in enhanced vibration which keeps the molecules well apart. In fact, they have so much space that they are free to rotate within the crystal lattice. The decrease in entropy corresponding to the specific heat anomaly is due to an ordering of this rotational motion of the hydrogen molecules.

Since then low temperature anomalies in the specific heat have been found in numerous substances. Sometimes we know the causative process but for many of them one is still in the dark about the mechanism involved. However, in each case it is quite clear that the entropy of the substance was higher than had been anticipated. It was this fact which opened the way to a range of temperatures much lower than had been accessible with liquid helium.

8 Magnetic cooling

After the successful liquefaction of helium in 1908 Kamerlingh Onnes returned again and again to attempts to reach lower temperatures. The method which he employed was always the same; to reduce the vapour pressure of liquid helium further, using more efficient pumps and in greater number. At each attempt he edged nearer to the ultimate limit which can be reached in this manner but there was nothing else he could do. Helium is the gas with the lowest critical data and once it had been liquefied and the vapour above it had been exhausted to the limit of the pumping capacity, the story of gas liquefaction was ended. Half a century had passed since Cailletet and when Onnes died, in February 1926, it seemed that he had taken the ultimate possible step towards absolute zero.

However, only two months later, on the 9th April, Professor Latimer of the University of California read a paper before the American Chemical Society in which an entirely new method of cooling was described. It dealt with the idea of a young instructor from Canada, William Francis Giauque who proposed to reach temperatures well below those of liquid helium by a magnetic method. Giauque's full paper, describing his idea in detail was submitted for publication on the 17th December. It was almost the story of Cailletet and Pictet all over again because a few weeks earlier, on the 30th October, quite independently the same proposal was sent in to the *Annalen der Physik* by Peter Debye.

In order to understand the mechanism by which magnetic cooling can be effected something has first to be said about magnetic effects in the structure of matter. The close relation between electricity and magnetism was put into rigorous form by Maxwell's electrodynamic theory, formulated in the second half of the nineteenth century. According to it a magnetic field is always connected with the motion, and in particular with the rotation, of an electric charge. The electrons which form part of the atomic structure are such moving charges and they are associated with magnetic fields in two ways. One is by orbiting around the nucleus and the other by rotation about their own axes, the spin. It is the spins which are of particular interest in low temperature phenomena.

At temperatures at which the spins are still disordered, they point randomly in all directions. When the substance is now brought into a magnetic field, say between the poles of a strong magnet, the lines of force tend to align the spins in the direction of the field. The degree to which the substance becomes magnetised by a given field is called its susceptibility. The alignment of spins by an external magnetic field is counteracted by the atomic vibrations which tend to disorder the spins. It is therefore easier, with a field of a certain strength, to align the spins when the thermal vibrations of the atoms are less violent; i.e. at low temperatures. This fact had been discovered at the turn of the century by Pierre Curie who noted that the magnetic susceptibility is inversely proportional to the absolute temperature.

Not all substances behave in this simple manner. In fact, in most of them the spins interact strongly with each other and, far from pointing at random in all directions, they pair off like two bar magnets which stick together with opposite poles. However, there are some crystals, in particular salts of the rare earth and iron group metals, that have a structure in which single spins are very isolated. These obey Curie's law and it was one of them, gadolinium sulphate, which Onnes and Woltjer had investigated in 1924. The main object of their experiment, which they achieved, was to see whether at a temperature as low as 1 K, and using a very strong magnetic field, they could approach complete alignment of the spins. They had chosen gadolinium sulphate because they had found that it still obeyed Curie's law, even at 1 K.

It was the full significance of this last fact which they failed to appreciate and which formed the starting point of Giauque's and Debye's considerations. As long as a substance obeys Curie's law, the spins must be in a state of disorder, pointing at random in all directions and this means that its entropy must still be high. At 1 K the thermal vibrations of the atoms in the crystal lattice of gadolinium sulphate have practically died out and the entropy due to them is negligibly small. However, the spin system of the salt is still disordered and may not pass into a state of low entropy until some much lower temperature is reached. The salt, like the solid hydrogen mentioned in the last chapter, is one of those substances which have still to undergo an important change below 1 K. On the other hand, a state of order can be forced on the salt's spin system at 1 K by placing it into a high magnetic field and it is the decrease in entropy achieved in this manner which Giauque and Debye proposed to utilise.

In principle the magnetic cooling method follows much the same steps as Cailletet's liquefaction of oxygen by first compressing the gas and then allowing it to expand. The only difference is, that instead of a gas, a salt is used and a magnetic field instead of the pressure. The simplest way to explain the magnetic method is by using a diagram in which the entropy of the salt is plotted against the absolute temperature (figure 8·1).

We have chosen the temperature region from somewhere above 1 K down to absolute zero. The heavy full curve denotes the entropy of the salt in zero magnetic field. At 1 K and for a good way below it the entropy changes little with temperature which means that it is entirely due to the spins which remain at random orientations in this range. Towards higher temperatures the entropy rises because now the lattice entropy becomes noticeable. The lattice entropy has been plotted separately as a dotted curve and this is parallel to the total entropy showing that the entropy of the spins remains unchanged. At some very low temperature the spins, even without the influence of any magnetic field must become ordered, complying with the third law of thermodynamics. This will happen when the thermal energy becomes so small that it is outweighed by the mutual interaction of the spins. This effect makes itself felt in our diagram by a sharp drop of the entropy (the full curve) to a vanishingly low value.

Next we introduce into the diagram the curves of the total entropy when the salt is placed into an external magnetic field. Three of these – dashed – curves are given, each corresponding to a field of different strength. Since the magnetic field has the effect of

8·1 The entropy diagram of a paramagnetic salt shows how very low temperatures can be reached by the magnetic cooling method.

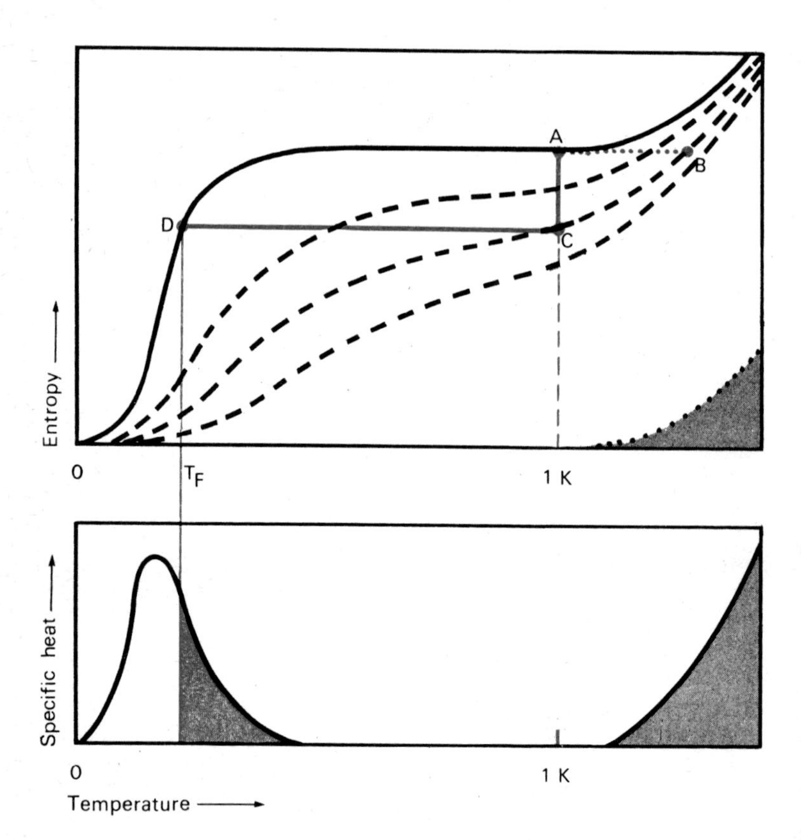

aligning the spins, they become more ordered than they were in zero field and our dashed curves therefore lie all below the full one. The entropy of the crystal lattice remains, of course, unaffected by the magnetic field and so its influence is the same in all three dashed curves. On the other hand, as demanded by Curie's law, the magnetisation, i.e. the spin alignment, increases with falling temperature. Consequently, the relative reduction of entropy below that in zero field becomes more pronounced as absolute zero is approached. The diagram is now complete except for sketching into it the operation of magnetic cooling.

Using liquid helium as a coolant, the temperature of the salt has been reduced to 1 K and no magnetic field has as yet been applied. It therefore occupies the position A in the entropy diagram. Magnetising the salt without allowing heat to enter or leave it is called an 'adiabatic' process the essence of which is that the entropy remains unchanged. Using a field with the strength of 1 tesla we can plot the effect of magnetisation by looking in our diagram for the point on the dashed curve for 1 tesla which has the same entropy value as A. This is B and we see that by magnetising it we would increase the temperature of the salt. Earlier we have compared magnetic cooling with the expansion of a gas. Before expansion, the gas has to be compressed and the warming of the salt from A to B is due to a heat of magnetisation which is quite analogous to the heat of compression, such as can be noticed by a bicycle pump getting hot. In an expansion cycle the heat of compression is carried away by the cooling water and in the magnetic process this is done by liquid helium. Provision is therefore made in the process to carry away the heat of magnetisation at 1 K and the salt is kept at this temperature, not, however, at A but on the curve corresponding to 1 T; and that is at C. This is now the starting point for the actual cooling step.

For the cooling step it is most important that the process should be carried out adiabatically, which means that any heat influx into the salt must be made as small as possible. The thermal connection with the liquid helium at 1 K is therefore broken and the salt is completely insulated. Then the magnetic field is reduced to zero and now we must find that point in the diagram which is on the full curve for zero field and has the same entropy value as C. This is D, which corresponds to a final temperature T_f much below 1 K. Magnetic cooling, as Giauque and Debye pointed out, could be achieved in this way.

After this first suggestion another seven years passed until the necessary equipment had been assembled to try out the method. The Leiden laboratory with its vast cryogenic facilities was still in the forefront but it was no longer there alone. In fact, in magnetic cooling Leiden ran a close second. Their first experiments were published in a communication dated the 15th May 1933, but Giauque at Berkeley in California recorded three successful experiments on the 19th March and the 8th and 9th April. The temperatures reached in these runs were 0·53 K, 0·34 K and 0·25 K respectively. The method had been proved beyond doubt and a new range of temperatures had been opened. After Berkeley and Leiden, Oxford and then Cambridge soon followed and by now magnetic cooling has become a standard method in low temperature laboratories all over the world. Improvements in technique, higher magnetic fields and more suitable salts soon extended the range of the new cooling method to temperatures below 0·01 K.

From our entropy diagram it may seem that, for any given salt there is not much point in using very high demagnetising fields because the entropy drop at the lowest temperatures is so very rapid. A much higher field would therefore be unlikely to achieve an appreciably lower temperature. This is indeed true but reaching a low temperature in itself is only winning half the battle. The other half is to stay long enough at the low temperature to make experiments. It is the same as in the case of gas liquefaction. The low temperature achieved by Cailletet's expansion of oxygen persisted only for a few seconds until the mist of droplets had disappeared. The ideal of having liquid gases boiling quietly in a test tube could only be attained when considerable quantities of these gases had been liquefied, so that they would not be evaporated again by any small influx of heat. In order to stay for a reasonable time at temperatures below 1 K not only the heat influx has to be minimised but an appreciable degree of spin alignment has to be achieved.

In our story we have come across the fact that there are usually two ways of explaining the same thing in our kind of physics. One is the rigorous thermodynamic approach which yields clear results but tells us little about the detail. The other is the microscopic theory which tries to interpret the happenings in terms of events on the atomic scale. The picture provided by the latter usually gives a better illustration of the phenomenon but is not always equally reliable. The explanation of magnetic cooling which we based on the entropy diagram was the thermodynamic one. The microscopic picture may make it easier to understand what has to be done to stay a long time at the low temperatures reached by magnetic cooling.

Let us first consider the case in zero external field. Although at 1 K the spins in a salt such as gadolinium sulphate are still disordered, they must, in accordance with Nernst's theorem, align themselves into an ordered pattern as absolute zero is approached. One of the ways in which this is achieved, and the only one which we need consider here, is by their magnetic action upon each other. We can think of them as little bar magnets, each with its north and south pole, which we indicate (figure 8·2) by little arrows. At 1 K, they point randomly in all directions. As the temperature is lowered and the thermal motion which jostles them decreases, the north and south pole of each magnet acting on the poles of its neighbours tends to pull all the magnets into the same direction. As the temperature is lowered further this spontaneous alignment grows until it finally is complete.

Looking at the same process but starting at very low temperatures we would find that at first all the spins are aligned. As heat is added the salt warms up but an extra supply of energy is now required to disorder the spin orientations. This energy is expressed as the specific heat of the salt which therefore in this region has to be high. Once the disordering has been accomplished no further extra energy is needed and the specific heat ceases to be anomalously high. Going back to the entropy diagram we therefore conclude that in the region where the entropy of the salt shows the rapid drop its specific heat has a peak. It should be mentioned that we could, of course, have arrived at exactly the same result

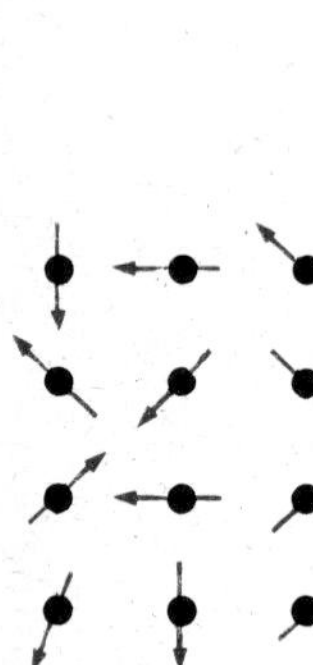

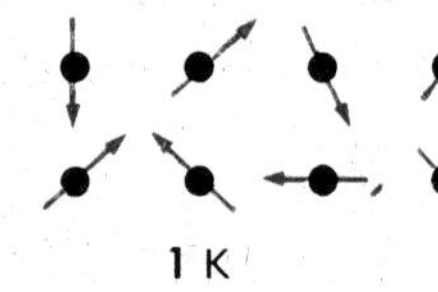

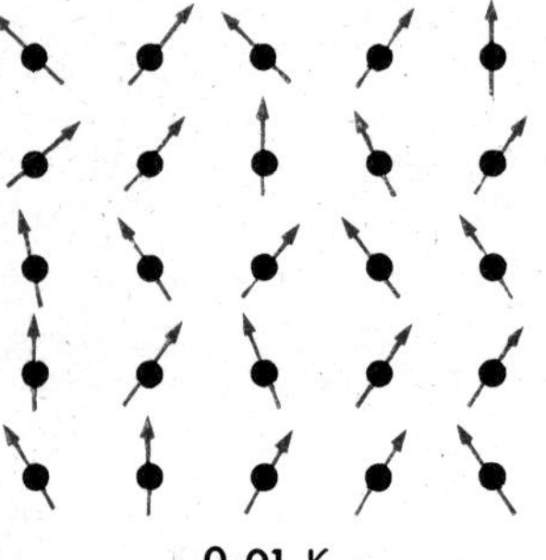

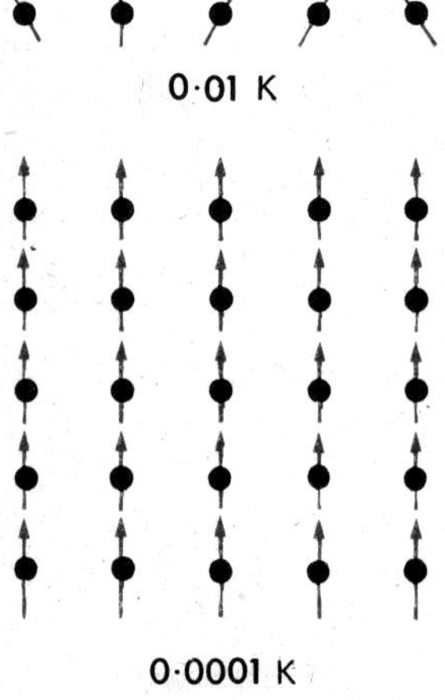

8·2 The increase in magnetic order as the temperature is lowered is shown in the direction of the electron spins.

by thermodynamics, except that this would have told us little about what is happening to the spins.

The unavoidable heat influx into a Dewar vessel filled with liquid helium results in the evaporation of some of the liquid. In order to stay at the low temperature we must have enough liquid in the vessel. Heat flowing into a magnetically cooled salt has the effect of disordering the aligned spins and in order to maintain its low temperature we must have sufficient spin alignment in the first place. In other words, we must have a high enough heat capacity at the low temperature if we want to keep the salt cold for a reasonably long time. Thus, a high demagnetising field may not bring the salt to a much lower temperature than a weak field but it will cause a much greater spin alignment. This is the same as to say that demagnetisation from higher fields will carry the salt deeper into the specific heat anomaly. The heat capacity at low temperatures is thereby increased.

Giauque's and Debye's proposal of magnetic cooling broke the deadlock which had barred the approach to absolute zero for nearly twenty years, after the liquefaction of helium. In theory the road to an uncharted new territory of 'magnetic temperatures' now lay open but whether it was a practicable road nobody knew. Without knowing how close they had been to the solution of the problem, Kamerlingh Onnes and Woltjer had provided the working substance by their researches on gadolinium sulphate. Four things were now required and had to function well at the same time: a low starting temperature, a strong magnet, a heat switch and, above all, thermal insulation much superior to any ever used before.

Kamerlingh Onnes had shown that it was perfectly feasible to maintain a temperature of 1 K or slightly below for many hours. He had achieved this by pumping off the vapour above liquid helium in a small Dewar vessel. This vessel was surrounded by others, containing in turn liquid helium at its boiling point, liquid hydrogen and liquid air. In order to achieve a high magnetic field, the pole pieces of a strong electromagnet have to be brought together as closely as possible. A pole gap wide enough to hold a bulky cryostat of a number of concentric vessels will not yield a field high enough for the experiment and cryostats of special design were required. The making of a high magnetic field is in itself an expensive item. Strong electromagnets, capable of creating a field of, say, one tesla over a sizeable volume require a lot of high quality iron which makes them not only costly but also cumbersome. The alternative is to pass a very high current through a solenoid, an iron-free coil, and to remove the heat generated by the current with oil or water cooling at a high rate. This poses tricky engineering problems and, in addition, a power station to generate the direct current needed for the experiment is required. Both these installations, iron magnets and solenoids, have been used for magnetic cooling.

The heat switch is needed to carry off the heat of magnetisation from the salt into the liquid helium at 1 K and then to insulate it from the liquid helium when the magnetic field is removed. The most commonly used solution is that provided by Dewar in his demonstration of the vacuum vessel mentioned in chapter 3. He showed to his audience a double walled vessel in which liquid air was boiling quietly because the space between the two glass walls was evacuated. When he broke the vacuum, the liquid began to boil violently and evaporated rapidly. Atmospheric air had entered the space between the walls and provided a thermally conducting path between room temperature and the liquid air. The same principle in a slightly more sophisticated form had been employed by Nernst in his specific heat measurements and it had become standard technique in low temperature research. Instead of breaking a seal and letting atmospheric air rush into the vacuum space, a tap is employed through which a small amount of gas, usually helium, can be admitted and through which it can also be pumped out when the vacuum insulation is to be re-established. Finally, good heat insulation has to be achieved by a very high vacuum and suitable construction of the cryostat in which all heat radiation into the experimental space has been excluded by reflecting shields.

When Giauque and Debye published their first papers on the

possibility of magnetic cooling in 1926, Leiden had ceased to be the only laboratory in the world where liquid helium was available. New low temperature laboratories began to be established in various countries and now a race for the new temperature range below 1 K was on which in a way resembled that for the first helium liquefaction. Even so, it took another seven years until the first successful magnetic cooling was achieved.

As is most fitting, the race was won, as mentioned above, by one of the originators of the scheme, Giauque. Debye, being a theoretician, was, in any case, out of it. On the 12th April 1933, Giauque, in collaboration with MacDougall, reported his first series of three experiments, carried out at the University of California. He had used gadolinium sulphate and, demagnetising his sample at a starting temperature of 3·4 K had, on the 19th March reached 0·53 K. Encouraged by this initial success, he improved the rate at which the liquid helium could be pumped off from his cryostat and on the 8th April, demagnetising from 2 K he achieved 0·34 K. An even better run on the following day started at 1·5 K and led to a final temperature of 0·25 K. Magnetic cooling had become reality and, already in these first pioneer experiments the lowest temperature obtainable with liquid helium had been reduced more than threefold.

This was just the beginning. Only a month later the Leiden laboratory, now named the Kamerlingh Onnes Laboratory in honour of its founder, reported the first successful cooling in which a temperature of 0·27 K was reached. The Leiden experiment had been carried out with cerium fluoride, another fairly expensive salt. However, when a year later the newly established low temperature laboratory at Oxford began its work on magnetic cooling, iron ammonium alum was chosen, a very common substance which is widely used to stop bleeding after razor cuts. Another feature of these experiments was that for the first time a second substance was cooled. Particles of cadmium were mixed intimately with the salt and it could be shown that this metal becomes superconductive at 0·56 K. A few years later demagnetisation experiments were

8·3 The four successive stages in a magnetic cooling experiment.

begun in Cambridge and after the war a large number of laboratories all over the world entered this field. While at first the main emphasis was placed on an investigation of the salts used in the cooling process, research on the properties of other substances which had been cooled to magnetic temperatures increased gradually. In the 1950s magnetic cooling had become a standard technique which presented no major difficulties, opening up a range of temperatures down to a few thousandths of a degree above absolute zero.

The procedure adopted for these experiments is much the same as that used in the first pioneer work at Berkeley, Leiden and Oxford. A schematic diagram is shown in figure 8·3. The salt *S*, usually in the shape of a sphere or an ellipsoid, is held by a support of low heat conduction in a container *P* which can be evacuated through the tap *T*. This container is surrounded by a bath of liquid helium, boiling under reduced pressure at about 1 K, in the Dewar vessel *D*. At first a small amount of helium gas is let into the container and the tap is kept closed. The salt therefore is also at 1 K because the helium gas is a good conductor of heat but the spins in the salt – shown diagrammatically as little arrows – remain still disordered at this temperature; their entropy is still high. Going back to our entropy diagram of figure 8·1, we can fix the position of the salt at this stage of the process as *A*. Next, a magnetic field is applied to the salt and this is usually done by bringing up a strong electromagnet so that the cryostat is now between the pole shoes (figure 8·3b). The external magnetic field forces the spins, which of course behave like little magnets, into the direction of its own lines of force. They become aligned in the field direction and thus present a pattern of order; their entropy is decreased and we will reach point *C* in the entropy diagram. However, as was explained earlier, the application of a field to the salt must result in a liberation of the heat of magnetisation and a tendency to move to point *B* in the entropy diagram. This warming up of the salt is being prevented by the presence of gas in the container *P* which conducts the heat of magnetisation into the liquid helium. In other words,

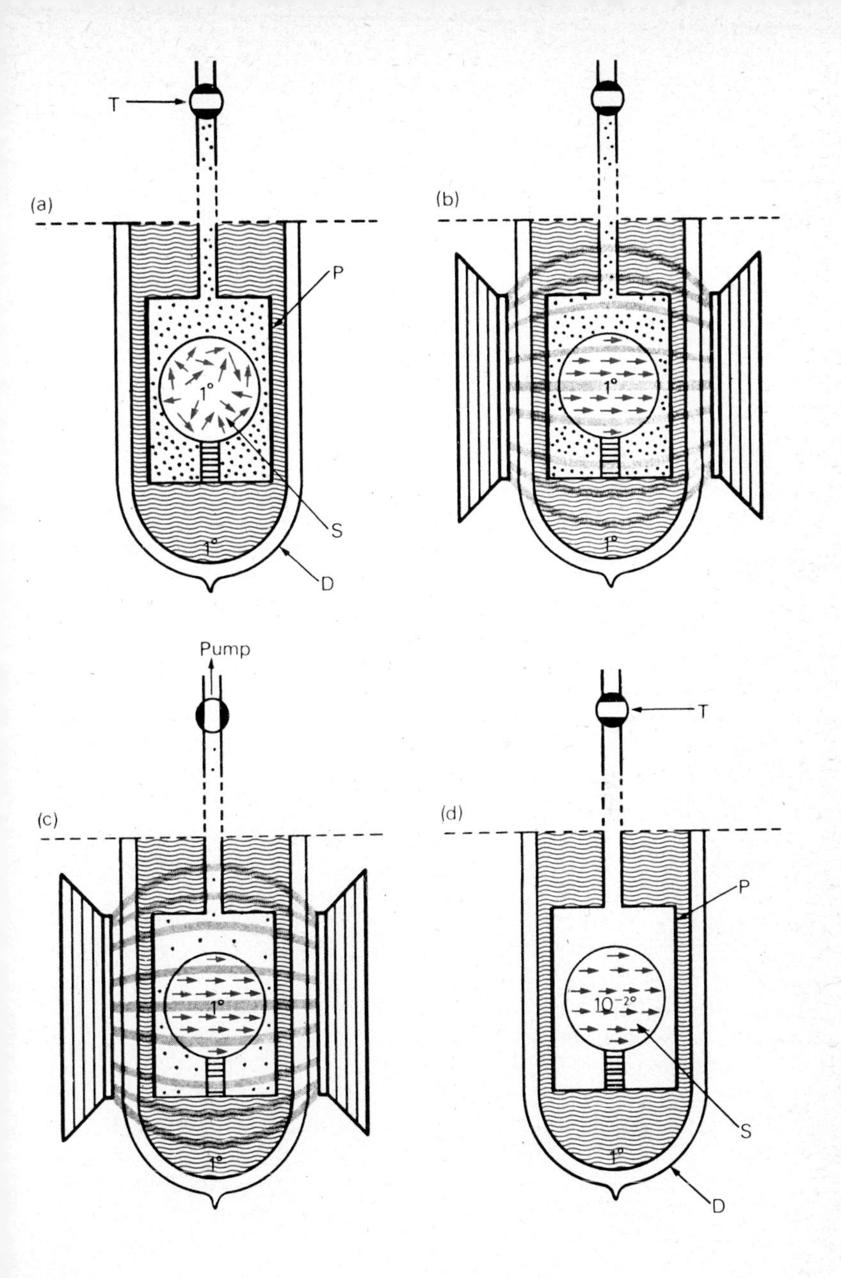
T
(a)
P
1°
S
1°
D
(b)
1°
1°
Pump
(c)
1°
1°
T
(d)
P
10^{-2}°
S
1°
D

the heat generated by the alignment of the spins merely results in some extra helium being evaporated but the temperature of the whole arrangement remains at 1 K.

The apparatus is now almost ready for the magnetic cooling to take place but before this can be attempted, provision has to be made so that no heat can flow into the cooled salt. The heat link with the rest of the apparatus has to be broken and this is achieved by opening tap T to the pump and evacuating the space inside P (figure 8·3c). Finally the electromagnet is taken away and the salt takes up position D in the entropy diagram which corresponds to the final low temperature reached by the process (figure 8·3d).

After this demagnetisation, experiments on either the salt or any substance cooled down by it can commence. The duration of time for which observations will be possible depends, of course, on the influx of heat into the salt. In his very first experiments already, Giauque had been able to maintain very low temperatures for several hours. How remarkable this achievement was became clear when in the first experiment at Leiden, one month later, it was found impossible to hold the low temperature for more than a few minutes. In the years which followed much work was done at the various laboratories to improve further the heat insulation of the salt and it has been possible to reduce the heat influx to 10^{-8} watts. This is indeed an extremely small flow of energy.

Some measure of how minute the quantities of heat are which matter in these experiments is provided by a strange disturbance that bedevilled the early experimenters for a long time. It was found that, on frequent occasions, heat seemed to enter the salt specimen from some unknown source. The heating was large enough to make accurate experiments impossible and had to be traced. The puzzling feature of the disturbance was that it took place sporadically and that in general it seemed to be worse in the daytime than at night. A final clue was provided by the fact that the heating seemed worse when mechanical pumps were operating in the vicinity of the cryostat. In the end it was tracked down to the fine nylon threads on which the salt had been suspended in order to reduce heat conduction. Mechanical shocks and running machinery, which occurred more frequently in daytime, were causing vibration of the nylon threads. The heat generated by the stretching of these threads when they vibrated was sufficient to spoil the experiment, and the disturbance vanished when rigid supports were used.

The ultimate limit of cooling which can be obtained by demagnetisation of a paramagnetic salt depends not only on the starting temperature and the magnetic field employed but above all on the magnetic properties of the coolant. Referring again to our entropy diagram (figure 8·1), we see that this limit is set by the steep drop of the entropy curve in zero field. We have also seen that this drop is due to the effect of the spins upon each other at a temperature which is so low that the thermal vibrations are too weak to prevent the spins from forming an orderly pattern. The problem is again very similar to the cooling of a gas in an expansion engine. There, too, as the temperature is reduced, the interaction of the cohesive forces will become predominant and the gas begins to liquefy. This necessarily means that further cooling by gas expansion becomes impossible.

The analogy between expansion cooling and magnetic cooling goes even further. An expansion engine using hydrogen will allow cooling to lower temperatures than one using air because the boiling point of hydrogen is much lower than that of air. Still lower temperatures can be obtained by a helium engine. It is the same thing with the paramagnetic salts, the interaction between the spins occurs at different temperatures for different substances. To give a few examples; the rapid drop in entropy takes place at about 0·2 K in gadolinium sulphate, at about 0·05 K in iron ammonium alum and at roughly 0·003 K in cerous magnesium nitrate. This latter value can be taken as the low limit of paramagnetic cooling.

When comparing magnetic cooling with expansion cooling, it should be remembered that the process discussed above is essentially similar to a single expansion stroke, such as was used by Cailletet. Somewhat complicated magnetic cooling devices which correspond to a reciprocating expansion engine have indeed been operated but their usefulness has turned out to be rather limited. The merit of such a magnetic engine lies only in its ability to cool down a large mass of a second substance. Since, however, at these very low temperatures the heat capacities of other substances are usually small compared with that of the salt, one single demagnetisation is in most cases quite sufficient. More than one demagnetisation is, however, useful when very low temperatures are to be obtained and for these purposes cooling devices with two, or even three, demagnetisation stages have been employed. In these cases different salts with successively lower temperatures for the ultimate entropy drop have been used in the individual stages. Here again we have a close analogy with the liquefaction 'cascades' described in chapter 2.

When Giauque and Debye first proposed their new cooling method, the question arose whether, and if so how, these very low temperatures could be measured. Fortunately this turned out to be a silly question, that is, one which already contains its own answer. A cooling can only be effected with a substance whose entropy still changes with temperature and the entropy change can, according to the second law of thermodynamics, always be used to determine the absolute temperature. In other words, any cooling mechanism provides automatically its own thermometer.

As with the magnetic cooling itself, its thermometry, too, is closely analogous to that used with gas liquefaction. Changes in the temperature of a gas result in a variation of its pressure and volume which are related to each other by the laws of Boyle and Gay-Lussac in the simple equation: $P \times V = \text{constant} \times T$. The corresponding relation for a paramagnetic salt is Curie's law which can be written as: Susceptibility = constant/T. It again is a very simple formula which suggests that magnetic thermometry also will be relatively simple. In the case of a gas thermometer the change of pressure of a given volume of gas is measured and this is then directly proportional to the absolute temperature. It is a very old method of temperature measurement which was first used by Galileo and by Amontons. In the case of a salt, all that is required

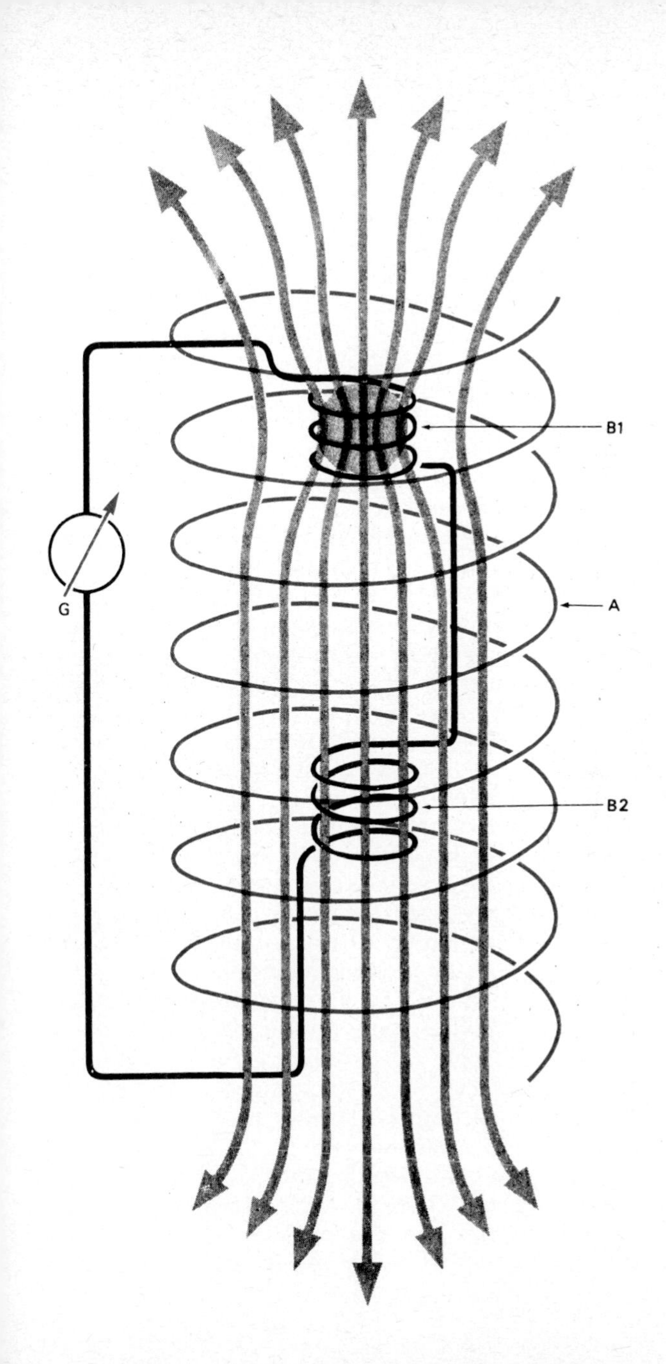
B1
A
G
B2

is to measure its magnetic susceptibility which, according to Curie's law is inversely proportional to the absolute temperature.

The measurement itself is relatively simple. The susceptibility is the magnetisation of the salt which is produced by a given magnetic field. In practice this is achieved by switching on a small magnetic field and measuring its strength inside the specimen. This can fortunately be done without placing any magnetic recording gear into the salt itself. When a current is switched on in a long cylindrical coil A (figure 8·4), a homogeneous magnetic field is established inside it and the strength of this field is measured by the current induced in the secondary coil B and recorded through the galvanometer G. The secondary coil is divided into two equal parts B_1 and B_2 wound in opposite directions. Thus, if the magnetic field had the same strength throughout the length of the primary coil A, no net current would flow in the secondary and the galvanometer would not show a reading. In order to act as a magnetic thermometer the system of coils is placed around the cryostat in such a position that the salt S comes into the centre of B_1. Since the salt has a paramagnetic susceptibility, magnetic lines of force are drawn into it, and as a result of this, their total number passing through B_1 will therefore be larger than that threading B_2. Because now the current induced in B_1 is larger and outbalances that induced in B_2, the galvanometer records it. This reading is a direct measure of the susceptibility of the salt and, according to Curie's law, of the reciprocal of the absolute temperature.

This simple determination of the magnetic temperature becomes less straightforward as absolute zero is approached and again the difficulties are similar to those experienced with the gas thermometer. We have seen in earlier chapters that, when a gas is cooled, there begin to appear at a certain temperature deviations from the simple equation $PV = RT$ which are due to the cohesive forces between the gas molecules and which herald the impending liquefaction of the gas. The interaction between the spins in the paramagnetic salt plays a similar part in causing deviations from Curie's law. This does not mean that the magnetic thermometer now becomes useless. It still will show a reading but a more complicated formula than Curie's law, one which contains corrections for the spin interaction, has to be applied. It is much the same thing as the substitution of the van der Waals equation for the simple gas laws.

However, application of these corrections is quite a tricky thing because it involves detailed knowledge of the complex magnetic interactions in the salt, and it becomes trickier the lower the temperature is which has to be measured. From this it would seem that the true value of the lowest temperatures which can be reached with the paramagnetic salts might always be shrouded in uncertainty. But fortunately we are again saved by the second law of thermodynamics. This law contains the definition of the absolute temperature itself. As was shown by Lord Kelvin more than a century ago, it is to be obtained from the cycle performed with an ideal heat engine, first considered by Sadi Carnot. Such an engine will do work by taking up a quantity of heat Q_1 at an absolute temperature T_1 and discard the heat $-Q_2$ at a lower temperature T_2. These quantities of heat and the absolute temperatures stand in the relation: $Q_1/T_1 = -Q_2/T_2$. This means, of course, that if Q_1, Q_2 and T_1 are known, the low temperature T_2 can be determined from this relation. As usual, the beauty of thermodynamic equations is that they are generally valid and do not only refer to any particular physical system. All that is demanded in our case is that no heat other than that accounted for in the relation shall enter or leave the apparatus while the measurement is performed.

In the case of the magnetic temperatures this last condition can easily be fulfilled. Since in any case extreme care is taken to reduce heat influx into the salt, the thermodynamic cycle to be performed for the thermometric measurement is very close to an 'ideal' one. There are a number of different ways in which this experiment can be carried out which differ only in technicalities which we don't have to discuss in detail. They are all based on the same principle which consists in measuring quantities of heat and relating them to one known absolute temperature in the established range. This, is, of course, T_1, the higher temperature, which in our case is the starting point for the demagnetisation, located in the range of liquid helium. The quantities of heat to be applied to the salt are given to it either by an electric heater or by irradiation with γ-rays and they can be measured out with very great accuracy.

The thermodynamic determination of the absolute temperature allows us to be completely sure in the mapping out of the new range below 1 K which is being explored, but it is a somewhat cumbersome procedure which one would not wish to repeat in each individual experiment. What is being done instead, is to carry out, once for each of the salts used, a very careful thermodynamic measurement. The results are then used for determining the correction terms mentioned earlier. This allows one to employ in all other researches with the same salt the simple technique of measuring only the magnetic susceptibility, and all that has to be done besides reading the galvanometer is to look up the corrections in a table prepared from the thermodynamic determination. In order to be able to make this comparison between different experiments, the geometrical shape of the salt specimen must be simple and this is the reason why it is usually made in the form of a sphere or an ellipsoid.

As mentioned earlier on, most of the researches carried out below 1 K have been devoted to investigation of the properties of the salts themselves. By using the thermodynamic temperature scale it has not only been possible to measure the true magnetic susceptibilities but also the specific heats. The most interesting features to be studied are the mechanisms through which, at the lowest temperatures, the interaction of the spins takes place. As absolute zero is approached, Curie's law gradually begins to fail and the magnetic susceptibility ceases to be proportional to the absolute temperature. Finally the susceptibility changes rapidly and usually in a very complex manner, indicating that the system of spins undergoes a profound change. This change, in accordance with Nernst's theorem, always signifies the establishment of a regular pattern in the spin system which, though not similar to it, never-

8·5 Specific heat of iron ammonium alum indicating two different mechanisms of spin ordering.

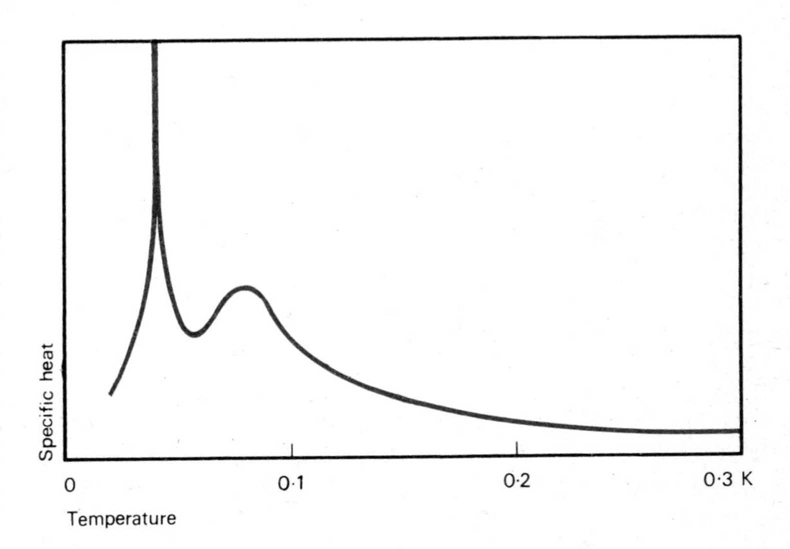

theless corresponds to the ordering of atoms in a crystal lattice. Only, for the spins it is not order in positions but order in directions. The main objects of research are the detailed nature of this pattern as well as the manner in which it is established. To give just two examples of the latter, the spins may align themselves through direct magnetic forces exerted upon each other, like a set of compass needles brought close together, or they may be aligned by the electric field of the atomic lattice. In the first case the effect is of a co-operative nature, relying on the mutual interaction of many spins upon each other, while under the influence of the crystal field each spin is aligned separately but all in the same direction. These changes in the degree of orderliness are changes of the entropy and they must make themselves felt in the specific heat of the salt. Measurement of the specific heat can, in fact, be used in order to tell us which type of alignment takes place in a particular salt. Co-operative effects tend to take place rapidly as the temperature is changed, the mutual interaction causes something like an avalanche in the change between order and disorder, and we must expect a sharp peak in the specific heat. On the other hand, the action of the lattice field on the individual spins will result in a more gradual change of the specific heat. In some salts both of these alignment processes will take place, though at different temperatures. As an example the specific heat of iron ammonium alum is shown in figure 8·5. As the salt is cooled a gradual ordering of the spins under the influence of the crystal field begins to be noticeable at about 0·2 K, leading to a broad maximum at 0·09 K. This is followed by a sharp peak, indicating co-operative ordering at 0·04 K. Measurements of the susceptibility have shown that below 0·04 K the salt now exhibits magnetic hysteresis, somewhat similar to that shown by iron at room temperature.

Most of the researches on a second substance cooled to below 1 K by means of the demagnetisation of a salt have been concerned with a continuation of observations made above 1 K. For instance, it was mentioned at the end of the last chapter that solid hydrogen shows a strong anomaly in its specific heat which was still found to rise at the lowest temperatures which could then be obtained. Magnetic cooling now provided a possibility of following up this phenomenon to much lower temperatures and of tracing it out as a quite unexpected curve, showing a large peak. Similarly magnetic cooling has revealed superconductivity in many more metals as well as new features of the strange behaviour of liquid helium which will be discussed in chapter 10.

In one respect magnetic cooling has led to an entirely new phenomenon; the orientation of atomic nuclei. The field which has been opened by these experiments belongs to the realm of nuclear physics and, in a way, low temperature technique has here only taken the part of a handmaiden to nuclear research. However, the results obtained are sufficiently important that we should give a short account of this work; particularly since it is closely connected with the attainment of still lower temperatures.

Just as the electrons, so the nuclear particles, too, the protons and the neutrons, have spin which confers a sense of rotation about its own axis to the atomic nucleus. It was to be expected from other observations and from theoretical considerations that the radiation emitted from a radioactive nucleus would occur in definite directions in respect to this spin axis. Since under normal conditions the spin axes of the individual nuclei point at random in all directions, no net directional effect will be observed in the radiation. If, on the other hand, the axes of the nuclei can be made to point all in the same direction, then the radiation, too, will be emitted in well defined directions from the radioactive sample. Such an alignment of the nuclei can be obtained by making use of the magnetic effect of the spin, its magnetic moment. Owing to the small size of the nuclear particles their magnetic moment is about a thousand times smaller than that of the electrons and that makes it so much more difficult to align them, even at the lowest obtainable temperatures. In fact at 0·01 K an external magnetic field of about five tesla is needed to produce nuclear orientation.

In spite of the formidable experimental difficulties inherent in this method it has been applied with at least some success. We have to remember that the salt used as coolant, in order to reach 0·01 K must be carefully shielded from the very high magnetic field employed for nuclear orientation so as not to warm up again. Fortunately there exist some very elegant tricks through which nuclear orientation can be achieved more easily than with 'brute force', as the method mentioned above has been nicknamed. One of these tricks which is the easiest to explain relies on the magnetic field of the electron spins. From our figure 8·4, illustrating the action of the magnetic thermometer, it can be seen that the paramagnetic salt attracts the magnetic lines of force. This means that inside it the magnetic field is higher than the applied external field. These internal fields can be quite strong, for instance at 0·5 K an external field of only 1/10 tesla will produce inside the salt a field of no less than 50 tesla.

The experiment is carried out with a salt whose ions have radio-

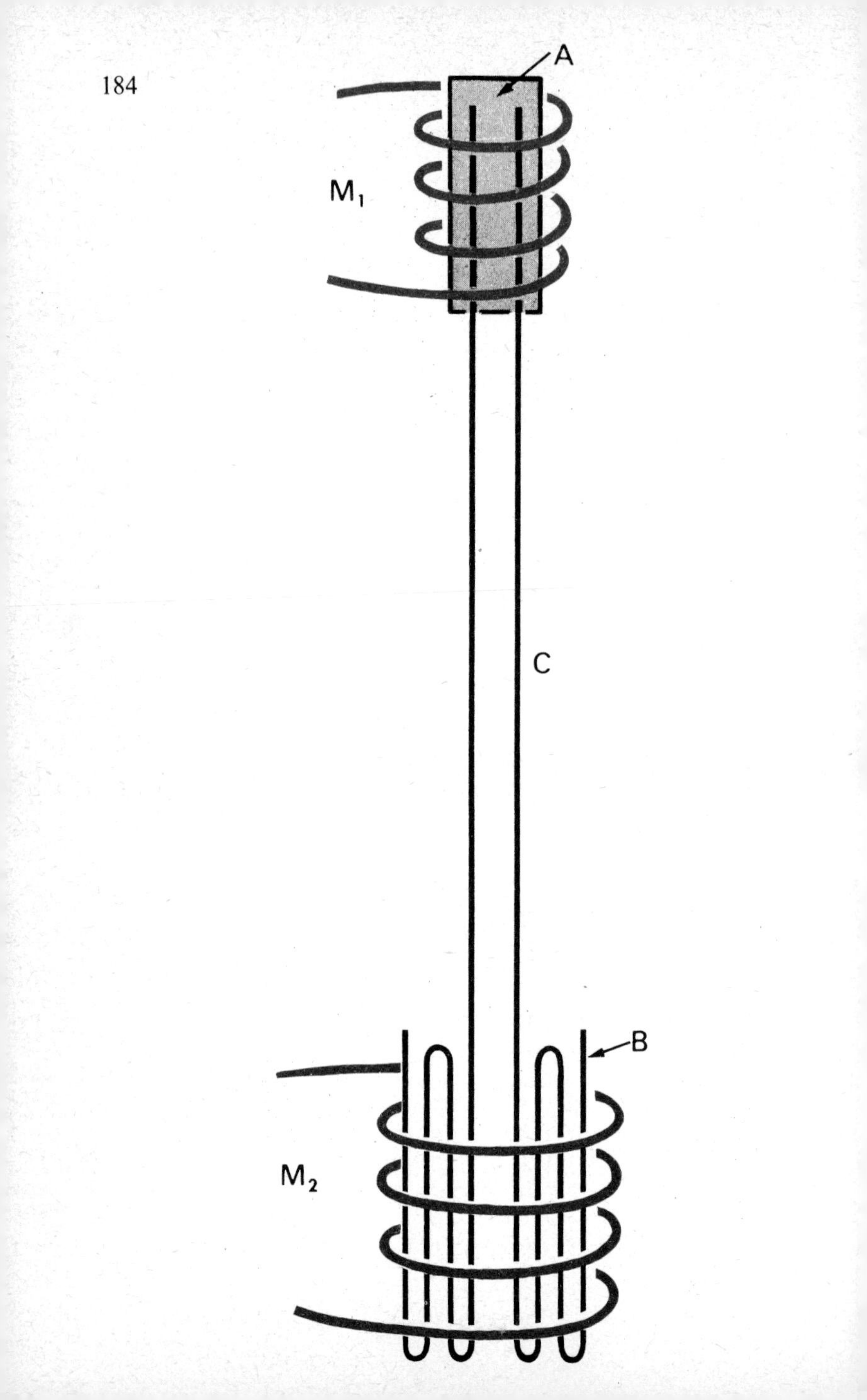
A
M1
C
B
M2

8·6 Nuclear cooling; (*left*) diagram of the apparatus and (*right*) the result. The curve gives the observed deflection of the magnetic thermometer which has been calibrated in absolute temperatures.

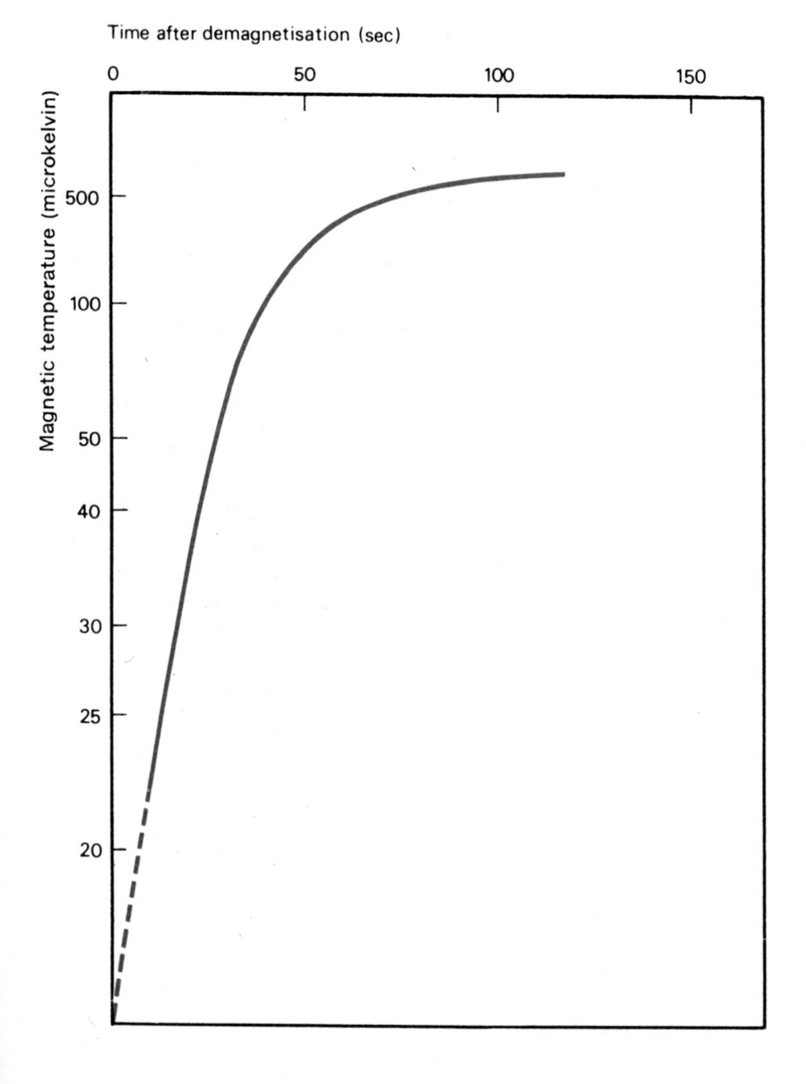

active nuclei. Demagnetisation from a high field starts, as usual, at about 1 K, but then the external field is not reduced to zero. Instead, demagnetisation is stopped at 1/10 tesla and now the spin axes of the radioactive nuclei are all oriented in the same direction; that of the external field. Counters for the detection of nuclear radiations are now moved through various positions in relation to the sample and the direction of the radiation coming from it is recorded. Among the important results obtained in this manner was the proof of non-conservation of parity in weak interaction, a theory for which Yang and Lee received the Nobel Prize in 1957.

The existence of paramagnetism due to nuclear spins had been discovered in 1936 by two Russian physicists, Shubnikov and Lazarev, working in Kharkov. It is obvious that with the successful cooling by means of electron spins, further cooling to still lower temperatures by demagnetisation of nuclear spins should have been considered. In fact, the feasibility of nuclear cooling was first discussed in the mid-thirties by Gorter in Leiden and by Kurti and Simon in Oxford. In principle the method is simple enough, being nothing else but a repetition, at lower temperatures, of what had already been accomplished with the electron spins in paramagnetic salts. However, its practical application poses enormous difficulties which are due to the smallness of the nuclear spins. The great difficulties just mentioned in connection with the brute force method of nuclear orientation are enhanced many-fold when nuclear cooling is to be achieved. Not only has the nuclear working substance to be cooled to about 0·01 K by means of a demagnetised paramagnetic salt but then it has to be isolated from it thermally and itself demagnetised from a field of at least 5 tesla.

With the limited means available for research in the late 'thirties there could be little hope of carrying through such an ambitious project, but in the late 'forties things began to look more hopeful. Great efforts were made by the late Sir Francis Simon at Oxford to assemble the necessary equipment and he, together with Kurti and their co-workers set out on the long road of preparing for the experiment. Then, in 1956, shortly before his death Simon had the satisfaction of seeing the first successful experiment. The temperature recorded amounted to less than 0·000016 K, less than two hundred thousandths of a degree above absolute zero.

This incredibly low temperature could only be attained for an instant and after about a minute the nuclear coolant had again warmed up to the temperature at which the demagnetisation had been started. Even so, in retrospect the success of the experiment appears almost like a miracle. Only gradually has it become clear that apart from the foreseen difficulties a host of further obstacles stand in the way of nuclear cooling.

The principle employed in the 1956 experiment is very similar to the two-stage cooling cascade which has been successfully operated with different paramagnetic salt in each stage. However, there was a great difference in degree. Now the first stage by itself had to be as efficient as the former double-stage cascade and then the nuclear stage had to be demagnetised from an extremely high magnetic field. With even the large available magnet facilities strained to the utmost, it was decided in the pioneer experiment to do without a heat switch between the two stages. A schematic drawing of the experiment is given in figure 8·6. *A* is the paramagnetic salt, representing the first magnetic cooling stage. It is connected with the nuclear coolant, *B*, through a link *C*. M_1 and M_2 are two water-cooled coils for producing high magnetic fields. Neither the cryostat nor the other ancillary equipment are shown. Metallic copper was chosen both for the nuclear coolant, *B*, and for the link, *C*. In fact, a large number of very fine copper wires were used. Their top ends were pressed into the paramagnetic salt, *A*, and their bottom ends were folded over several times to form *B*. This arrangement had the advantages of good heat contact between copper and salt and of avoiding eddy current heating on demagnetisation since the copper was in finely divided form.

First the electron spin stage *A* was demagnetised and by heat conduction through *C* the nuclear stage *B* was cooled to a temperature of about 0·02 K. Then the nuclear stage was demagnetised from the highest field which could be produced, which was just under three tesla. A few seconds after this field had been reduced to zero, measurement of the nuclear magnetic susceptibility was begun and the readings obtained are shown in figure 8·6, in which the temperatures calculated from Curie's law are also given. As already mentioned, the temperature of the nuclear stage rises rapidly after the demagnetisation so that after about one minute most of the nuclear cooling is lost again. We see from the diagram that when recording of the nuclear susceptibility was started, a few seconds after the field was switched off, the temperature recorded was 0·000022 K, and by working back to the beginning of the experiment, the maximum cooling to 0·000016 K can be deduced.

At first it was thought that the rapid rise in temperature was merely due to the fact that a heat switch in the link *C* had been omitted and that heat flowing from the salt along the copper wires into the nuclear stage was warming up the latter so quickly. It therefore seemed that the introduction of an efficient heat switch, such as had already been used successfully in two-stage demagnetisation of paramagnetic salts, would remedy the situation. It was only when experiments in this direction were undertaken that the real difficulties of nuclear cooling became apparent. These obstacles are of a much deeper nature than any technical inadequacy and they are somewhat discouraging.

In order to understand all this, we must have a closer look at the concept of temperature. When discussing in earlier chapters a gas we described its temperature in terms of the kinetic energy of the molecules and in a solid by the vibration of the atoms. In addition, we have seen that the susceptibility of the spins, too, is a measure of temperature. In the copper used for the nuclear cooling stage all three of these phenomena are present. There are the vibrations of the atoms, the nuclear spins and also the kinetic energy of the degenerate electron gas and we have tacitly assumed that all three should, at any time, give the same reading for the temperature. At normal temperatures, and even at those of liquid helium, this is true, but only because the exchange of energy between spins, electrons and vibrations of the crystal lattice is instantaneous. Experiments with paramagnetic salts below 1 K have shown that, while the energy exchange is not quite as rapid, it is generally still fast enough not to falsify the determination of temperature. Things change, however, very much when still lower temperatures are considered.

At 0·00002 K energy exchange of the nuclear spins with the electrons and the lattice vibrations is a very slow process when compared with the time of observation, and this leads to a quite different interpretation of the 1956 experiment than had at first been believed. On demagnetisation the nuclear spins follow immediately the change of the external field and take up a state which corresponds to 0·000016 K but the electrons and the lattice retain the starting temperature. One is thus faced with the strange situation of the copper sample having simultaneously two different temperatures; 0·000016 K for the spins and 0·02 K for electrons and lattice. Then the gradual energy exchange between the spins and the rest of the substance warms up the spins to the starting temperature again.

In spite of many efforts and a great deal of ingenious experimentation, it was not until more than ten years after the first nuclear demagnetisation that the electrons and the lattice could be cooled effectively to temperatures of the order of 0·05 K and held there for several hours. On the other hand much information has been gathered on the behaviour of the demagnetised spin system and it has become customary to make a distinction between *nuclear cooling* which refers to a lowering of the spin temperature only and *nuclear refrigeration* which denotes the decrease in temperature also of a second system, such as the electrons and lattice or even another substance, say liquid helium. It is the latter process which is of far more general importance and which has played a significant part in the investigation of the properties of matter in the millikelvin range.

Two steps were essential for the successful development of nuclear refrigeration. The first was an effective heat switch and the second a more efficient primary cooling stage. Both these features were needed for cutting down the heat leak into the nuclear stage, thus giving time for the nuclear stage to absorb the heat discarded by electrons and lattice. In fact, as will be apparent from what has been said above, allowing sufficient time for this energy exchange is the crucial factor in operating any scheme or research, other than on the nuclear spins only, at temperatures of 0·001 K or below.

A heat switch can be provided by a short length of superconductive wire which is interposed between the primary cooling stage and the demagnetised nuclei. Rendered normal by a small magnetic field the wire provides excellent heat flow between the two stages which is decreased a million-fold when it is made superconductive by removing the field. Paramagnetic salts as primary coolants have been replaced by the helium dilution refrigerator. This new cryogenic method, which we shall discuss in the last chapter, can provide temperatures approaching 0·01 K in continuous operation and is thus superior to the use of paramagnetic precooling. This combination of heat switch and the dilution refrigerator for nuclear refrigeration has been used in the last few years by a number of laboratories and the present record is held by Lounasmaa's research group at Otaniemi, Finland. He has been able to reduce the electron and lattice temperature in a copper specimen to 0·0003 K and maintain it, and the same group has cooled liquid helium to 0·0007 K. This is an impressive achievement permitting observations on the light helium isotope in its most interesting temperature range.

9 Superconductivity

Whatever surprises the new and only partly investigated region of magnetic cooling may have in store for us, they can hardly exceed those which have already been revealed by the temperature range of liquid helium. Kamerlingh Onnes had the first glimpse of the strange new world of superfluids when in 1911 he discovered superconductivity. In more than half a century which has passed since then our knowledge of these enigmatic phenomena has greatly increased and we have also come much nearer to a theoretical interpretation. Even so, we are still far from understanding the true meaning of what evidently is a fundamentally new aspect of aggregate matter. As we have just seen, the existence of such unusual phenomena as the zero point energy or gas degeneracy could be predicted by theory but we have to admit that, even with all the knowledge we have today, we could never have predicted the superfluids. With no prediction or theory to guide us, the exploration of the superfluids has been a search without landmarks, a blindfolded groping for new information without direction. Because of this large degree of uncertainty in the search for the pieces of the puzzle, its history has provided the most exciting chapter in the quest for absolute zero.

Kamerlingh Onnes was a genius but one given to sober thinking and so it took a while before he began to realise the enormity of his discovery. As is only natural, his first impulse was to look for a connection between superconductivity and the known physical phenomena. For a little while he hoped that the drop in the resistivity of mercury was something which he had half expected, but he soon had to realise that unwittingly he had stumbled upon the unknown. Such vague theories of the electrical resistivity as existed at the time might have been coerced to allow for a fairly rapid decrease with temperature, and this indeed was tried by Onnes, but he had to give up when it turned out that the resistance disappeared suddenly and discontinuously.

Faced with what he now recognised as an entirely new state of matter, Onnes set to work exploring it. His very first experiments had been designed to find out how small the resistance of mercury

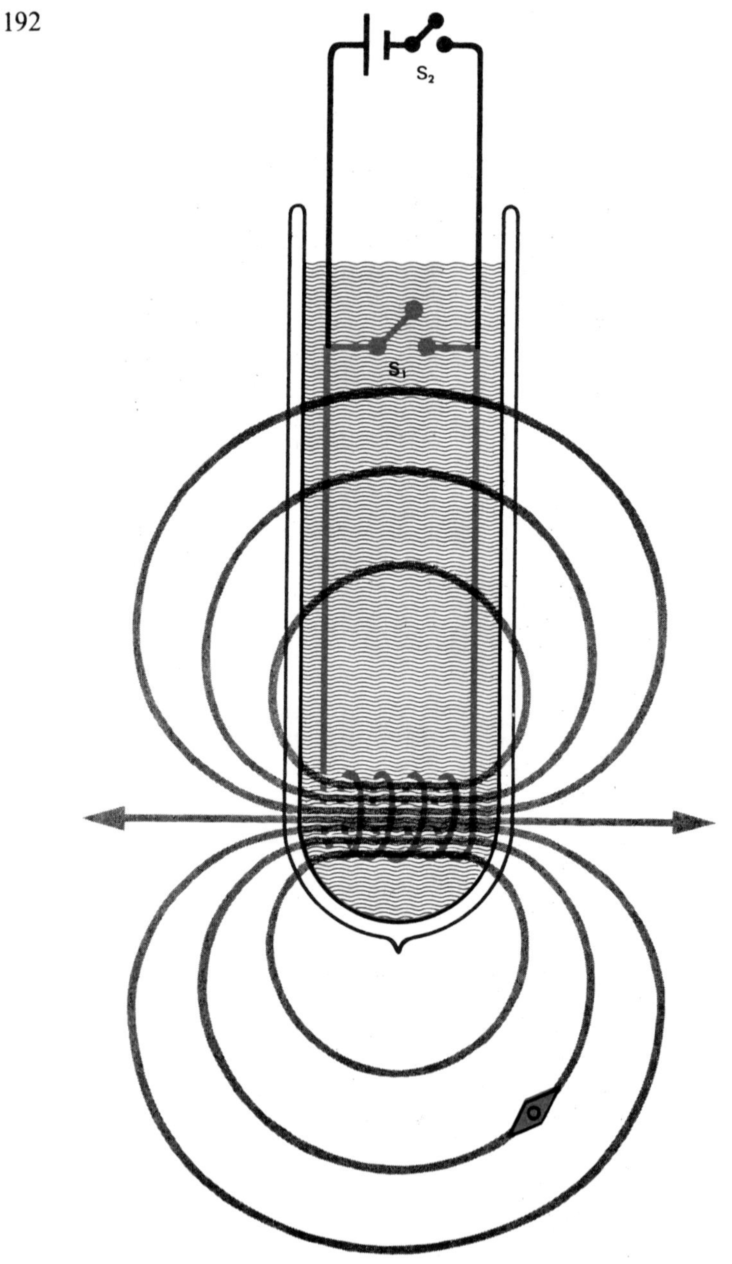
S2
S1

9·1 Kamerlingh Onnes' demonstration of zero electrical resistance in a superconductor.

had become and now he had to ask himself whether it had perhaps vanished completely. This immediately led to an awkward question of principle. It is always possible in physics to find out ultimately if a quantity is very large or very small; it is never possible to say that it is infinite or zero. The best that can be done is to state that it is larger or smaller than a limiting value derived from experiment. What was required in the case of superconductivity was to improve the accuracy with which a very small resistance could be detected, far beyond the limit set by the conventional instruments. Here Onnes was in his element. Within three years he had devised an ingenious method which has not been superseded to this day.

He made a coil of superconductive lead wire as shown in figure 9·1 which could be short-circuited by a cleverly designed superconductive switch S_1. With S_1 open, current was fed into the coil from a battery at room temperature through normally conducting copper wires. This current could be cut off by opening the switch S_2. The experiment began by closing S_2 but leaving S_1 open. Under these conditions current from the battery was passing through the superconductive coil where it produced a magnetic field which could be detected by the deflection of a compass needle outside the Dewar vessel. Then S_1 was closed and S_2 opened which means that the superconductive coil was now short circuited while at the same time the current supply from the battery was cut off. However, the compass needle still showed the same deflection as before, indicating that the current was still running through the coil, although it had ceased to receive energy from the battery. The reason for this strange phenomenon lies in the lack of electrical resistance of the coil which permits the current to pass through the wholly superconductive circuit without dissipation of energy. This 'persistent' current, as it came to be called, provided Kamerlingh Onnes with an extremely sensitive method for detecting any small trace of resistance which might remain in the superconductive lead coil. If energy is dissipated in it, the current must gradually decrease as time goes on and this would be recorded by a change in the deflection of the magnet needle. Until, after several hours, the

9·2 The lead sphere floats in space, supported by persistent currents.

liquid helium in the Dewar vessel eventually evaporated and the coil ceased to be superconductive, not the slightest decrease in the deflection of the compass needle had been noted; the persistent current had remained unchanged. From this experiment Onnes could conclude that any resistance of the superconductive lead coil must be at least a hundred thousand million times smaller than at room temperature.

Shortly afterwards Onnes performed the same experiment in a still simpler way. He used a single closed ring of lead which he cooled in the field produced with a magnet outside the Dewar vessel by filling the vessel with liquid helium. When now the magnet is taken away, the magnetic lines of force cannot escape out of the ring because it is superconductive and any alteration of the magnetic flux will be compensated for by the induction of a persistent current in the ring. At the end of the operation Onnes was therefore left with the ring carrying a current and having trapped a bundle of magnetic lines of force. The existence of this persistent current can again be noted on the deflection of a compass needle.

A very impressive demonstration of persistent currents is shown in figure 9·2 where currents have been induced in two such rings. A lead sphere is then dropped into these rings but, as it approaches them, persistent currents are now also induced in the surface of the sphere through the magnetic field generated by the rings. These currents run in the same direction as those in the rings and, through their magnetic fields they repel each other with the result that finally the sphere will float in space above the rings at a distance at which this magnetic repulsion has become equal to the weight of the sphere.

An equally spectacular effect is provided by a small bar magnet which is lowered on a chain into a dish of superconductive lead (figure 9·3). As the magnet approaches the dish we notice that the chain becomes slack and eventually we see the magnet hovering above the dish. What happens here is that the lines of force from the magnet induce persistent currents in the lead surface which

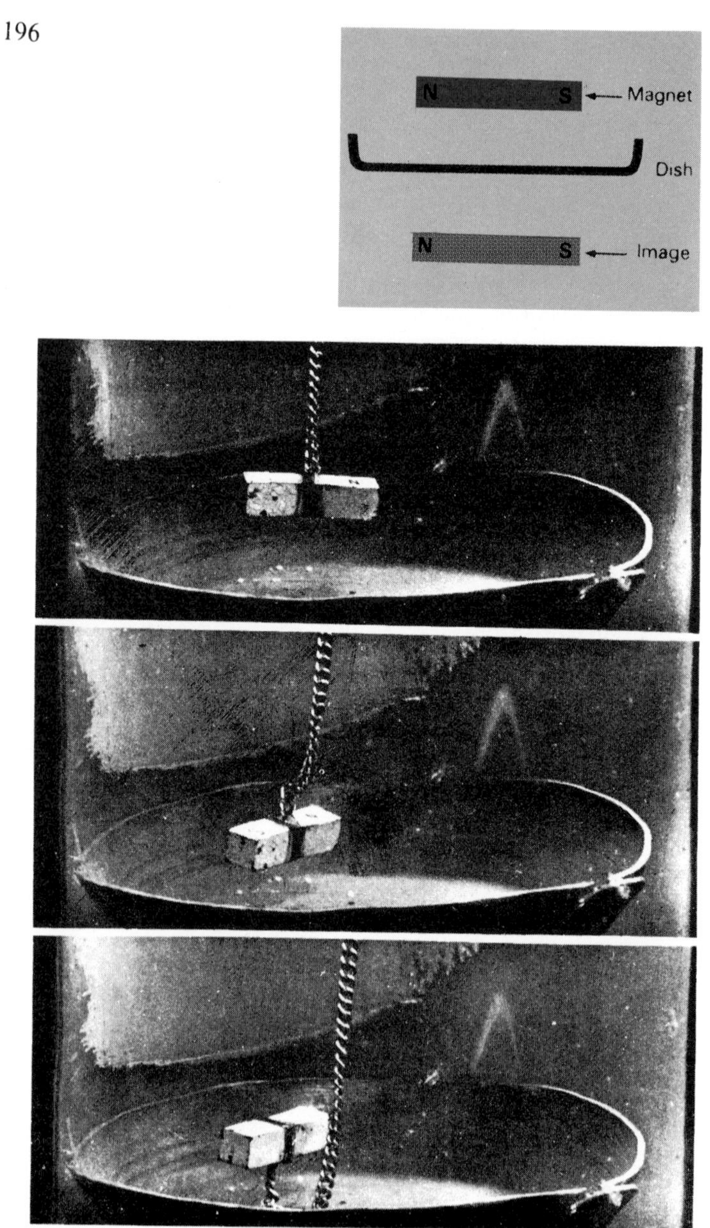
N
S
Magnet
Dish
N
S
Image

9·3 A bar magnet dropped into a superconductive dish is repelled by its own magnetic image below the dish.

compensate exactly the field of the magnet. The little bar magnet therefore 'sees' its own magnetic image which lies below the lead surface at the same distance at which the real magnet stands above it. The north and south poles of the image repel those of the real magnet which therefore cannot approach the dish further.

Kamerlingh Onnes's experiments with persistent currents have since been repeated, first by him and then by others, with more sensitive detecting devices and by prolonging the duration of the experiment. The longest time for which a persistent current has been kept running was about two years and it would still be running today if a transport strike had not cut off the supply of liquid helium. Even after two years there was no sign of any weakening of the current and we are therefore not too far off the truth when we treat a superconductor as having zero resistivity.

One of the first things which occurred to Kamerlingh Onnes was the use of a superconductive coil to produce a very high magnetic field. When discussing magnetic cooling the great technical difficulty of creating strong magnetic fields was mentioned. The large currents required for this purpose generate enormous quantities of heat in the magnet coils which has to be removed by cooling water, the circulation of which sets a limit to what can be achieved. The prospect of using lead coils, as that in figure 9·1, which operate without dissipation seemed to offer magnificent possibilities. Loss-free superconductive transformers as well as other forms of electrical machinery without resistance provided tantalising prospects. Admittedly, huge quantities of liquid helium would be required, but this price would not be too high for the saving in energy that could be achieved.

Unfortunately, the dream was short-lived. Research in Leiden soon showed that superconductivity breaks down when a certain critical magnetic field is exceeded. It was also found that this 'magnetic threshold' was relatively low, never exceeding more than a few hundredths of a tesla, a value far below that required for even modest electrical machines. In spite of this discouraging revelation, the story of high field superconductivity did not end with it but its sequel had to wait for another forty years. In the meantime it was soon discovered that the thresholds of magnetic field and current are really one and the same thing because the critical field strength which can destroy superconductivity in a wire is just the same as that which is created by the critical current flowing through it.

All the time the search for new superconductors went on. In addition to mercury and lead, tin, indium, thallium and gallium were soon found to be superconductive. They all are metals with rather similar physical properties such as low melting points and softness. When in the 'twenties other laboratories began to join the search, Meissner in Berlin turned his attention to a different group of metals, those which are hard and have high melting points and he found amongst them a new series of superconductors such as tantalum, niobium, titanium and thorium. With the extension of observation to the neighbourhood of 1 K and particularly to the temperatures accessible by magnetic cooling a further crop of superconductive elements was found such as aluminium, cadmium, zinc, osmium, ruthenium and many others. In most cases the samples had not to be very pure to show the effect but in others superconductivity was only found recently when extremely pure material became available. As things stand now, the superconductors far outnumber the non-superconductive metals and it is a moot point whether those which have not shown any sign so far of becoming superconductive may do so at still lower temperatures.

This kind of doubt really concerns only some of the normal metals which are left, as for instance the monovalent ones such as gold or sodium and a few divalent metals like magnesium and calcium. Another group of non-superconductors as for instance iron, cobalt, nickel and the rare earth metals all have high internal magnetic fields which are likely to suppress superconductivity and they will therefore always remain normal.

It was hoped that the occurrence of superconductivity in certain metals might help in understanding superconductivity better altogether. At first it was believed that certain atoms might be superconductive and others not but this idea was soon proved wrong when superconductivity was discovered in the compound Au_2Bi although neither gold nor bismuth are superconductive. Moreover it turned out that ordinary white tin only is superconductive while the grey modification of the metal remains normal. Here the only difference is the structure of the crystal lattice which in the first case is tetragonal and in the other cubic. These facts showed that the reason for superconductivity had to be sought in the gas of free electrons rather than in the nature of the atom. Even today, when an enormous number of experiments have provided some guidance in the relation between crystal structure, the number of free electrons and superconductivity, the correlation is not really good enough to predict the occurrence of superconductivity.

Another pointer that superconductivity has its origin in some relatively subtle arrangement of the free electrons is provided by the fact that the phenomenon is confined to low temperatures. The energy involved in any transition, and also that to superconductivity, can quite generally be assessed again by the quantity kT, where T in this case is the transition temperature. Since no superconductor has been found much above 20 K, T, and with it the energy, is too small to be ascribed to some process occurring inside the atom. In fact only a few alloys show a transition temperature near 20 K while those of the pure metals have been found to all lie below 10 K.

With work on superconductivity spreading to the newly established low temperature laboratories and thus growing in volume, it had by 1933 become increasingly clear that some vital piece in our knowledge still seemed to be missing. What exactly was wrong could not immediately be seen but a number of clues pointed to the magnetic behaviour. Certain measurements carried out at that time in Leiden on the transition of wires in a magnetic field appeared to be in flat contradiction to some of their earlier results. Another inconsistency arose from Onnes's hope of making high magnetic fields. In 1930 de Haas and Voogd had found that wires of lead–bismuth alloy remained superconductive in magnetic fields as high as two tesla. When three years later I installed

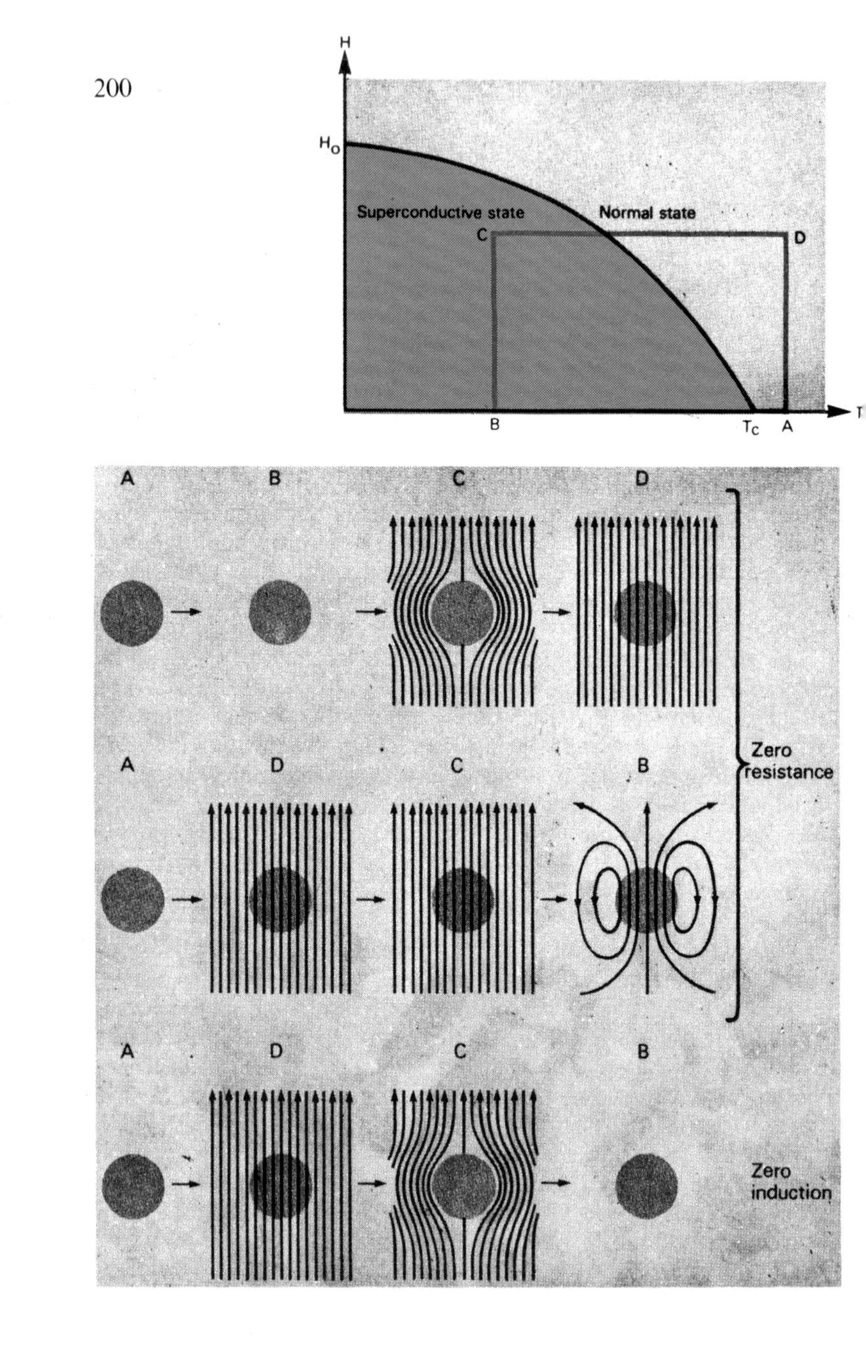
H
H_o
Superconductive state
Normal state
C
D
B
T_C
A
T
A
B
C
D
Zero resistance
A
D
C
B
A
D
C
B
Zero induction

the first helium liquefier in England, at the Clarendon Laboratory in Oxford, this alloy seemed to be just the right material with which to construct a superconductive magnet. A coil was made of it but it failed to produce the expected high field.

While we were still scratching our heads, trying to find out what had gone wrong with our magnet, Meissner and Ochsenfeld in Berlin announced a discovery which completely changed all ideas about superconductivity. In order to understand its meaning, we must have a closer look at the threshold curve which denotes the boundary of the superconductive state in the diagram of magnetic field plotted against temperature (figure 9·4). The area enclosed by this roughly parabolic curve is the region in which the metal is superconductive while at higher temperatures or magnetic fields, beyond the confines of the curve, it is normally conducting. At the transition point' T_c a vanishingly small field is already sufficient to quench superconductivity. This point varies, of course, from metal to metal, being for instance 4·1 K in mercury, 0·56 K in cadmium and 9 K in niobium. As the temperature is lowered below T_c the metal will retain its superconductivity even in the presence of a magnetic field and the maximum field H_0 which it can withstand at absolute zero again differs from substance to substance. For the series of metals just mentioned the values are 0·04, 0·003 and 0·17 tesla respectively.

Now, let us consider what will happen to the magnetic field in the neighbourhood of a metal sphere whose electrical resistance becomes zero as we cool it through T_c. Since the cooling (from A to B) is done in zero external field, no changes are to be expected. At B we increase the field around the sphere from zero to a value, C, well below the critical one. Since now the sphere is perfectly conducting, the rise of the external field is compensated for by super-currents in its surface and no magnetic flux can penetrate into it. Thus the magnetic lines of force are bulging around the equator of the sphere and this local increase in magnetic flux can easily be detected by a suitable instrument. When the sphere is now warmed up in a constant external field from C to D, the supercurrents in its surface die out as the threshold curve is passed and magnetic flux penetrates into it until, at *D*, the field inside and outside the sphere is the same.

An entirely different pattern, however, had to be expected when the sphere is passed through this same cycle in the opposite direction. Raising the field from *A* to *D*, the lines of force will penetrate it because it is normally conducting. On cooling in a constant field from *D* to *C* the sphere becomes perfectly conducting but since we do not vary the external field, no magnetic effect is to be expected. On the other hand, perfect conductivity will make itself felt in the magnetic behaviour of the sphere when now the field is decreased from *C* to zero at *B*. As the lines of force try to leave the metal, this magnetic change is compensated for by persistent currents being set up in the surface of the sphere which prevent the flux from escaping. We should thus be left at *B* with a magnetic dipole in the sphere which in its action resembles a bar magnet.

Until 1933, it was generally believed that in a superconductor this pattern of behaviour actually obtains and the belief seemed to be supported by an early experiment made at Leiden. Unfortunately, however, this sphere had been hollow, an important circumstance which appears to have escaped everybody's notice. Only when the experiment sketched in figure 9·4 was properly carried out on solid samples of metal by Meissner and Ochsenfeld did the true magnetic pattern of the superconductive state become apparent. The cycle *ABCD* yielded, in fact, the expected magnetic behaviour but when it was performed in the opposite sense *ADCB* a most surprising effect was discovered. In the section *DC* the magnetic flux did not remain in the metal but was completely expelled on passing through the threshold curve. Accordingly, on reducing the external field from *C* to *B*, the sphere retained no flux which could be trapped and no magnetic dipole was left at *B*.

The new effect discovered by Meissner is the spontaneous expulsion of magnetic flux from the metal when it becomes superconductive and the appearance of this phenomenon also results in a new symmetry of magnetic behaviour because it is evidently immaterial in which sense the cycle is performed. We have dealt with the actual experiment in some detail because only in this way can it be seen that zero electrical resistance by itself is insufficient to explain the Meissner effect. Thus, in addition to the disappearance of resistivity, the superconductive state also shows a magnetic effect; the disappearance of magnetic induction.

In the first year after the new discovery a number of efforts were made to connect the two effects, that of zero resistivity and that of zero induction with each other through Maxwell's electrodynamic equations. However, it soon became clear that all these attempts were doomed to failure because the relevant equations became indeterminate for this case. This showed that superconductivity cannot be comprehended in the framework of ordinary electro-dynamics and that it required an entirely new approach which was soon provided by Fritz and Heinz London who had come to Oxford as refugees from Nazi Germany. The starting point of their work arose from Heinz's doctoral thesis in which he considered the depth to which a persistent current penetrates into the superconductive surface and he had come to an interesting result to which we will return shortly. His own and his elder brother's arrival in England coincided with the discovery of the Meissner effect which lent fresh emphasis to his work, now aided by the superb theoretical skill of Fritz. At the same time this was a hectic period at the Clarendon Laboratory where more data on the Meissner effect and its significance were obtained by the experimentalists, leading to unending discussions between them and the theoreticians. As a result of this close collaboration, the Londons succeeded in formulating a new electro-dynamic framework for superconductors within less than two years after the discovery of the Meissner effect.

The central feature of the London electro-dynamics is a new equation connecting the current with the magnetic field which replaces Maxwell's well-known relation between current and electric field. Apart from providing a comprehensive description of the electromagnetic phenomena of superconductivity, the London equations reveal a satisfying symmetry of the superconductive pattern. In ordinary electro-dynamics a steady current is associated with a steady magnetic field but a steady magnetic field does not result in a current. In order to obtain a current in a normal conductor, the magnetic field must vary with time and this necessarily leads to asymmetry of the electro-dynamic equations. In the superconductor, on the other hand, the relation between steady current and steady magnetic field is, as the Londons showed, symmetrically balanced by the fact that a steady field will also cause a steady current. These are the persistent currents first demonstrated by Kamerlingh Onnes and the currents of the Meissner effect which make their appearance when the metal is cooled to superconductivity in a steady magnetic field.

The strange phenomenon of superconductivity thus really presents a much simpler pattern than normal electrical conduction, exhibiting a beautiful symmetry which is destroyed as we leave the vicinity of absolute zero. In fact, if Faraday had conducted his pioneer experiments at the Royal Institution with superconductive lead wires in liquid helium, instead of using copper wires at room temperature, he would have discovered laws of electromagnetic induction so clear and simple that he would have recognised in them a basic revelation. What we call normal conductivity might have appeared to him as a complex departure from them.

Another important consequence of the Meissner effect is provided by the expulsion of the magnetic flux from the metal when it is cooled through the threshold curve because it permits the application of thermodynamics. The pushing in and out of magnetic flux at the transition between C and D in figure 9·4 represents the performance of work, just as work is done by the expansion or compression of a gas enclosed in a cylinder. The thermodynamic relations are, as we have seen earlier, so general that it is quite immaterial whether the work is done against a piston or against the magnetic force. It is interesting to note that, even before the Meissner effect provided a firm basis, it was generally suspected that superconductivity must be amenable to thermodynamic treatment and its complete formalism had, in fact, been worked out by Gorter in Holland shortly before Meissner's discovery. Checking the prediction of these formulae was the next task for the experimentalists in Leiden, Oxford and Kharkov.

It was in this work that the first clue to the failure of our high field magnetic coil made of lead-bismuth alloy was found. While in pure metals the thermal effects, such as the specific heat and the latent heat of transition, agreed well with the values predicted by thermodynamics from the threshold curve, this was not the case for the alloys. Owing to the extremely high threshold values, very pronounced thermal effects could be expected in the alloys but work in Oxford as well as in Kharkov showed conclusively that there was no sign of these effects. Again it was clear that some important part in our knowledge of the superconductive state was missing and, in particular, that alloys seemed to behave quite differently from the pure metals.

With that much information to go on, the experimental approach was marked out pretty clearly in the direction of closer magnetic studies on superconductive alloys. The year 1934 was a busy one for the experimentalists in Oxford, Kharkov and Leiden, all teams working on much the same problems and using similar methods which, however, were sufficiently different to permit a useful check on each other's results. The upshot of all this activity was that, when a conference was held at the Royal Society in London in the spring of 1935, the whole pattern had emerged reasonably clearly. The alloys did in fact, behave quite differently from the pure metals. Instead of keeping the magnetic field out completely until the threshold field was reached, magnetic flux was found to penetrate the alloy samples at a relatively low field. However, when the lines of force entered the alloys there was not the same sudden destruction of the superconductive state as in the pure metals. The penetration of the metal by magnetic flux was found to be a gradual process, extending over a wide range of field, the specimen remaining electrically superconductive throughout this

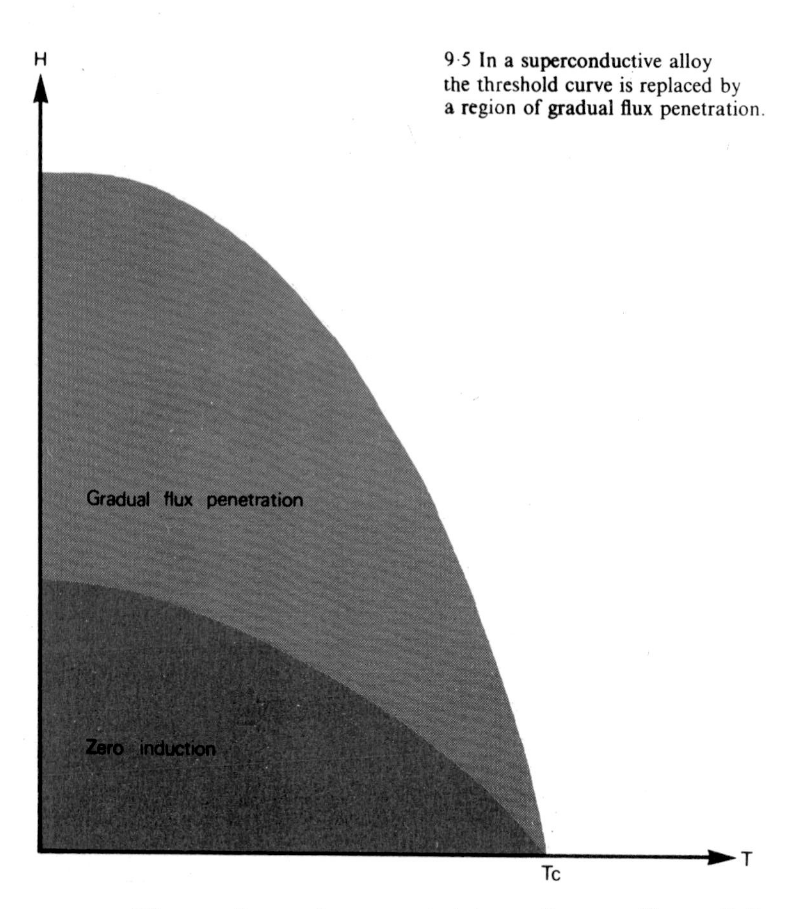

9·5 In a superconductive alloy the threshold curve is replaced by a region of gradual flux penetration.

process. The result can be presented in a diagram (figure 9·5) similar to that used for pure metals. The difference is that in an alloy the threshold curve is now replaced by a wide region of gradual flux penetration.

The significance of this curious phenomenon was realised only slowly and the fruits of this research were not gathered until thirty years later. Let us now go back to the subject of Heinz London's thesis which also was published in 1934. Our earlier statement that a magnetic field will induce persistent currents in the surface of a superconductor, while quite adequate to account for the phenomena which we have described, is nevertheless somewhat crude. A current is a flow of electrons and electrons repel

9·6 Penetration of magnetic flux into a thick and a thin superconductive wire. In the fine wire the magnetic induction is nowhere completely zero.

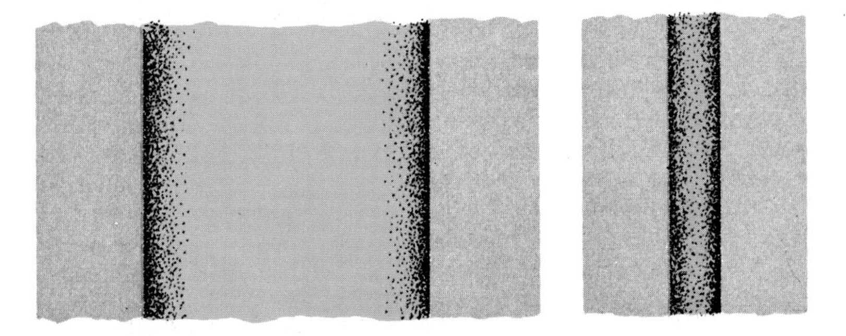

each other. We therefore cannot assume that the persistent current runs in the geometrical surface of the metal; it must extend to a certain distance into the metal. This distance was calculated by Heinz London who found that it will be about 10^{-5} centimetres and this means, of course, that the magnetic field, too, will penetrate into the metal to the same depth.

The proper superconductive volume of a metal sample will therefore be just a tiny bit smaller than its geometrical volume because the skin of 10^{-5} centimetres has to be subtracted. It is clear that this error is usually too small to be detected with specimens of ordinary size but it will, of course, become important when thin wires or films of metal are concerned (figure 9·6). Then a strange difficulty arises. We have seen that the work, the energy involved in the transition between the superconductive and the normal state of the metal, is given by the expulsion of the magnetic field from the volume of the sample. Since this is a basic thermodynamic quantity, it cannot depend on the shape of the specimen. Thus, if the metal is in the form of a fine wire, whose dimensions are not large in comparison with the penetration depth, the volume from which the magnetic flux is expelled is correspondingly smaller than in a bulk specimen. It follows that the only way to retain the same value for the energy of transition is for the threshold field to become correspondingly larger.

This conclusion on the basis of the London electrodynamics is of special interest because it provides a possibility of checking the theory by a direct experiment. There were certain difficulties. Techniques for making very fine wires were not so well advanced at the time and moreover, the metal had to be free from strain or impurities so as not to have alloy properties. Even so, the task was accomplished in 1937 by an American research student in Oxford, Rex Pontius, who had succeeded in making lead wires so fine that their diameter became comparable with the calculated penetration depth. The result was extremely gratifying in that not only an increase in the threshold field was observed but its magnitude was exactly as had been predicted by the theory.

Encouraging as this proof of the theory was, it nevertheless led to another awkward question. The experiment had shown unambiguously that a fine wire can retain its superconductivity to much higher fields than the same metal in bulk form. Why, we may ask, does the Meissner effect occur at all? Why does the metal, when placed in a magnetic field, not split up instead into an array of fine normal and superconductive regions each of which could then remain perfectly conducting up to very high magnetic fields?

Fortunately, the answer to this question was well known from another branch of physics; the behaviour of water drops. There, two or more small droplets, when coming into contact with each other, will always coalesce into one larger drop. The reverse process, a spontaneous splitting up of a single drop into a number of droplets does not take place. The reason for this phenomenon is, of course, the surface tension which leads to a gain of energy by reducing the total surface. A single spherical drop represents the minimum surface of a given quantity of water and no spontaneous break-up into droplets can occur because such a process would lead to a larger surface and thus require energy.

Applying the same thoughts to superconductivity, we must conclude that the existence of the Meissner effect indicates a positive surface tension at the boundary between the superconductive and normal states in the metal. Magnetic flux is expelled altogether from the sample because too much energy would be required for the formation of the great amount of surface between any fine superconductive and normal regions. At the same time these ideas provide us with an explanation for the strange behaviour of the superconductive alloys; they are apparently metals with a negative surface energy. With the realisation of this fact, the puzzling failure of the Oxford high field magnet became clear. It had been designed under the assumption that the alloy from which it was constructed behaved in the same way as a pure metal; and this assumption had been unjustified.

Research on superconductive alloys continued in a few laboratories until it was interrupted by the outbreak of war in 1939. It was not until the early fifties that the question of the negative surface tension was tackled again. Then Pippard in Cambridge investigated the nature of the interface between the superconductive and normal states of the same metal and concluded that in addition to Heinz London's penetration depth of 10^{-5} cm another quantity had to be considered which he called the 'coherence length'. The concept of coherence in the moments of the electrons in a superconductor was not a new one. It is implied in the observation of a persistent current and the phenomenon of zero resistance already indicates that in a superconductor momentum remains unaltered over a long distance while in the normal state the free path of an electron is interrupted by the next impurity which it encounters and at which it is scattered. Roughly speaking the coherence length is that distance below which it is impossible to distinguish between normal conduction and superconductivity. Even for the same substance its size must depend on the degree of purity. In very pure tin, for instance, an electron may travel for about 10^{-4} cm before it is scattered and within this space there is no difference to a superconductor. On the other hand, if the metal is impure or even an alloy the normal mean free path, and with it the coherence length, will be much shorter, say of the order of 10^{-6} cm or less.

Now, the surface tension at the boundary between the superconductive and normal states depends on the relative size of penetration depth and coherence length. If the latter exceeds the former, as is the case in a pure metal, the surface tension will be positive and the metal will keep out magnetic flux until at the critical field it will change discontinuously into the normal state. In an alloy the coherence length will be shorter than the penetration depth, the surface tension will be negative and flux can enter the metal at a comparatively low field. However, the normal resistance is not restored at this juncture but superconductivity is retained to very much higher fields. In short, the theory now accounted for the strange behaviour of alloys which we had observed twenty years earlier (figure 9·5).

In 1957 Abrikosov in Moscow published a full theory of this form of superconductivity, now called type II, predicting in some detail the behaviour of an 'ideal' superconductive alloy. He submitted his work at a conference in Moscow, the first one in which a small number of low temperature physicists from Oxford and Cambridge were present. It was an exciting occasion when at last we met personally the colleagues whose work we knew so well from their publications. One notable exception was Kapitza whom I had last seen almost a quarter of a century earlier when he left Cambridge to go on a visit to Moscow. Abrikosov postulated that on increasing the magnetic field around the alloy there should at first be a regime of zero induction which at a relatively low critical field (H_{c1}) would give way to a new 'mixed' state at which quantised lines of force entered the metal. On further increasing the field superconductivity should eventually vanish at a second, much higher, critical field (H_{c2}) beyond which the metal will be entirely normal (figures 9·5 and 9·7).

In recent years the existence of these quantised vortex lines of flux which, as Abrikosov predicted, are arranged in a regular lattice, has been proved by direct observations (figure 9·8). However, when in 1957 we listened to Abrikosov his ideas were not readily taken up. The trouble was that none of the magnetisation curves observed on alloys so far exhibited the predicted behaviour; instead of the sharp peak at H_{c1} there was an ill-defined maximum

9·7 Magnetisation curve of a pure type II superconductor showing the 'mixed' state between the fully superconductive and fully normal regions.

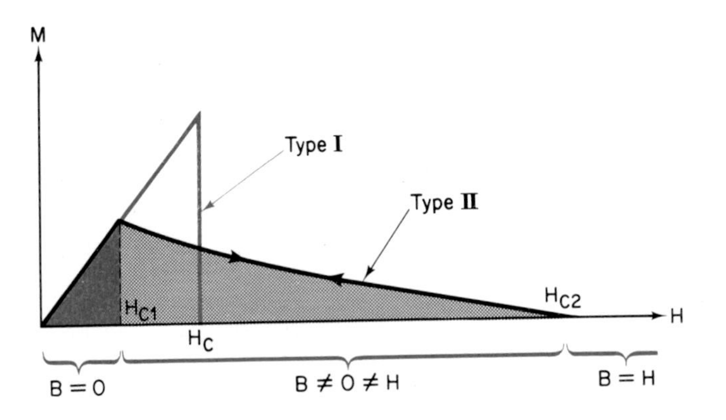

and none of the curves ever showed any sign of magnetic reversibility. It took some time until it was realised that Abrikosov's predictions only referred to an undisturbed single crystal of an extremely homogeneous alloy. No such specimen had ever been produced but, employing a recently developed technique, the Service Electronics Research Laboratory in Britain succeeded five years later in growing a tantalum–niobium single crystal which, when being measured at Oxford, exhibited in every detail the predicted behaviour of a type II superconductor. We shall return to the involved story of superconductive alloys and their importance for superconductive technology in the next chapter.

However, before becoming engrossed in technological considerations we must return to the basic problems of superconductivity. The most essential of these and the one which we have avoided so far is the question of an explanation.

As was inevitable, ever since superconductivity was first discovered many different theories for its explanation have been proposed; roughly at the rate of two or three per annum, and for the better part of half a century. New axioms designed to elucidate the phenomenon were formulated, ranging from the crudest to the highly sophisticated, which had all one thing in common; that

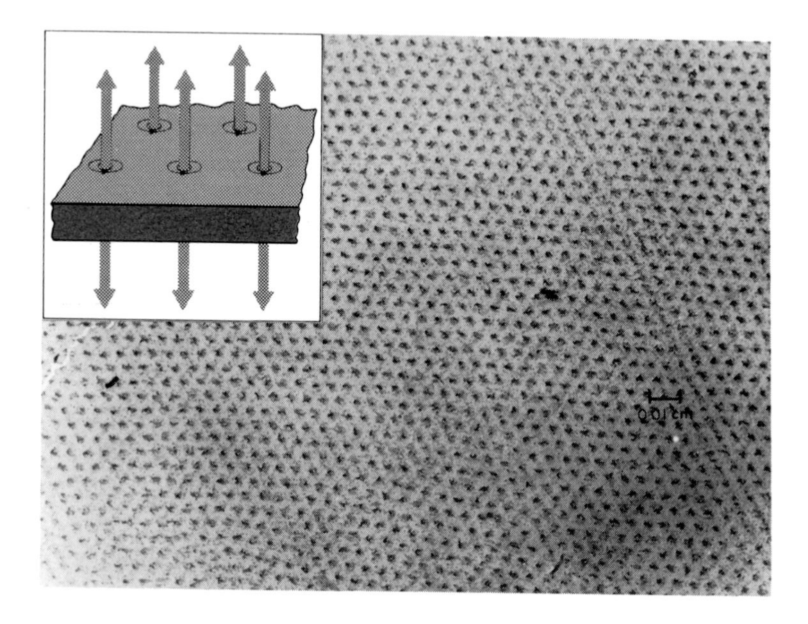

9·8 The regular array of quantised magnetic vortex lines in the mixed state can be made visible through the application of ferromagnetic dust to the surface of the metal.

they did not fit the facts. Eventually Felix Bloch who had done so much for our understanding of electrons in metals enunciated an axiom of his own which ran: 'Every theory of superconductivity can be proved wrong'. And for a long time this axiom turned out to be the only correct one.

However, very gradually the approach to the solution could be narrowed down by discarding a number of attempts as impossible and by learning more from experiments. The London electrodynamics and the application of thermodynamics did much to clarify the issues. For instance, from an experiment made at Oxford in 1938 it could be concluded that the entropy of a persistent current is zero. This discovery immediately suggested a small gap in the available energy states at the Fermi surface mentioned in chapter 7. This means that, as the temperature is

9·9 The energy gap at the Fermi surface of a superconductor.

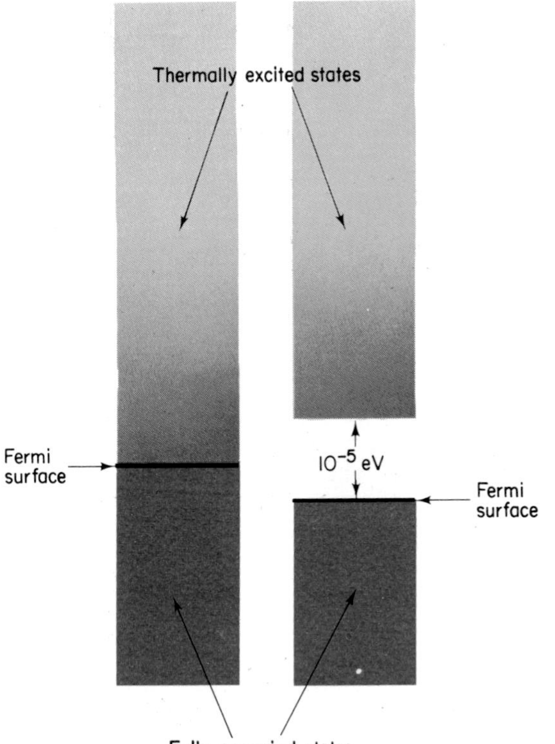

raised from absolute zero, the surface does not merely become fuzzy as in a normal metal, but that it remains smooth even at finite temperatures and that the electrons have to receive a definite amount of energy (about 10^{-4} eV) in order to jump over this gap before they can enter the excited, fuzzy states above it (figure 9·9).

This concept of an energy gap as the most significant feature of a superconductive metal explains why energy states below the gap cannot dissipate energy and must manifest themselves in resistance-free currents. The smallness of the gap also shows why superconductivity is confined to the lowest temperatures. The existence of the gap and its accurate measurement have since been established in a great variety of measurements. In this way the path for the theoreticians was mapped out step by step and all that remained to be done was to find some electronic process within the metal which would fit in with every one of these conditions as well as with the other known phenomena of superconductivity. It was a formidable task requiring a superb knowledge of electronic phenomena in metals, great mastery of mathematical technique and, above all, brilliant but controlled imagination.

What now appears to have been the crucial step was taken independently and at the same time, in 1950, by Fröhlich in Liverpool and Bardeen in Illinois. The basic idea of their theories is that the mechanism leading to superconductivity is an effect exerted by the electrons upon each other but through the intermediary of the lattice vibrations. Broadly speaking, and we cannot attempt here to give more than a very rough idea, superconductivity is due to the manner in which the atomic vibrations are influenced by the presence of the conduction electrons. Let us assume for simplicity's sake that we deal with a metal in which each atom has contributed one free electron. This means that the crystal consists of a lattice of positive ions, each an atom which has lost one electron, and a 'gas' of the same number of free electrons moving at random through this crystal lattice. Considering one negative electron in the lattice, surrounded by positive ions, the latter will be attracted to it and this slight local contraction of the crystal lattice will have two effects. The first is to cause an attraction for other electrons – and this can lead to superconductivity – and secondly it will be felt in the vibration of the positive ions. Taking these two effects together Fröhlich predicted that the onset of superconductivity should be affected by the mass of the vibrating ions.

Fortunately, the progress of nuclear technology has permitted the separation of isotopes, i.e. of atoms of the same element having different mass, of several of the superconductive metals. In fact, unknown to Fröhlich the experiment was already in progress in two American laboratories. When the results came to hand, they showed a small change in the transition temperature with ionic mass as predicted by him. Naturally, this result was considered extremely encouraging since it was the first time that a theoretical prediction of a superconductive phenomenon had succeeded.

However, for the theoretical interpretation of superconductivity, it was the end of the beginning rather than the beginning of the end. The experiment had proved that the type of interaction between electrons and lattice vibrations as suggested by Fröhlich was the relevant one but from there it was a long way before a particular mechanism belonging to this type could be shown as likely to cause superconductivity. Strangely, one of the troubles with the theory was that it was rather too successful and it was not clear why not all metals should readily become superconductive. Some important restricting condition was still missing and this was found in 1956 by Leon Cooper. Almost twenty years earlier Fritz London had invoked Bose–Einstein statistics to explain superfluidity in liquid helium which will be described in a later chapter. As we have seen in chapter 7, electrons in a metal obey Fermi–Dirac statistics in which owing to its spin each electron can occupy only one phase cell. On the other hand, in Bose–Einstein statistics where spins are paired, as for instance in a helium atom, many atoms can occupy one cell. Occasionally it had been suggested that such pairing might possibly take place also in the case of electrons and in some way cause superconductivity. Cooper now showed rigorously that the Fröhlich–Bardeen attraction mechanism would favour a kind of condensation by pairs of electrons at the Fermi surface having equal but opposite momentum and spin, and thereby lead to the energy gap. Moreover, attraction will only occur when the two electrons are not too close to each other, the optimal distance being the coherence length. In the following year Bardeen, Cooper and Schrieffer expanded this idea into a full theory of superconductivity for which they received the Nobel Prize.

Proof that the Cooper pairs are the carriers of a supercurrent was obtained in an unexpected manner. The persistence of such a current immediately suggests that it must be quantised, since otherwise it would diminish in strength and its energy be dissipated as radiation. The fact that this does not occur can only be explained by assuming that the spectrum of permitted energy states of a persistent current is not continuous but consists of steps each of a size of the elementary 'fluxoid', as postulated by F. London. Its value is extremely small, h/e (where e is the electron charge) which means only about 10^{-15} of the unit of magnetic flux. The experimental difficulty of detecting such minute steps in a magnetic field is formidable but in 1961 it was solved, independently by teams in California and Germany whose results were published in the same issue of *Physics Letters*. An apparent blemish in both investigations, the fact that the steps were only half the predicted size, was first ascribed to some inaccuracy of the observations until it was realised that the value was not h/e but $h/2e$. The carriers of a persistent current are not single electrons but electron pairs.

The central phenomenon of superconductivity, the complete loss of electrical resistance now finds its explanation in the behaviour of Cooper pairs. Normal resistivity is due to the scattering of single electrons which on collision change momentum. The Cooper pairs, too, suffer scatter but the scatter which one of the pairs undergoes is compensated for by equal and opposite scatter of the other electron; this means that the total momentum of the pair is conserved. It suffers no loss of momentum and the current of electron pairs flowing through the metal can thus do so without loss of energy, in other words: completely free of resistance. Resistance only reappears when, as the temperature is raised through the transition point, heat motion causes the pairs to be broken up.

When it gradually became clear that the Fröhlich–Bardeen mechanism was going to provide the key to superconductivity there was widespread disappointment among physicists that such a striking phenomenon should have revealed nothing more exciting than a footling small interaction between electrons and lattice vibrations. Clearly here we were witnessing for the first time the operation of quantum laws on a macroscopic scale and one expected the disclosure of some new fundamental principle of nature. However, such a new principle was indeed involved but it was so strange that its recognition took a long time to sink in. Understanding was made even more difficult by the mathematical form in which the theory had to be presented.

The new fundamental principle is pair formation and we are as yet far from comprehending its full significance and its place in the mechanism of natural phenomena. Its basic feature appears to be the maintenance of the momentum of the pairs which allows them to carry this momentum over long distances without dissipation through an environment which is classically scattering. The idea of connecting frictionless transport with pairs of opposite spin first arose in 1938 when Fritz London drew attention to a condensation not necessarily according to position, but in momentum space. London tried to apply this model to liquid helium and we shall leave the discussion of spin pairs to the final chapter dealing with liquid helium.

10 Technology near absolute zero

We have mentioned in the last chapter how Kamerlingh-Onnes's dream of huge loss-free magnets and other aspects of a new superconductive technology was shattered by the discovery of the limiting critical field. The short revival of hopes provided by the discovery of alloys that retained zero resistance in high magnetic fields was dashed by the unexplained failure of the Oxford coil and a similar attempt made a little later in Leiden. We will later in this chapter return to this problem and its solution but for the time being the discouraging results stopped all further progress along these lines until interest was revived by an accidental observation twenty years later.

While the use of strong currents and high fields had to be excluded from applications of superconductivity, this did not affect its possible use in devices employing weak currents and small fields. These, as was to be expected, were in the domain of instruments rather than machinery and they had the added advantage of requiring little liquid helium, then still a somewhat rare commodity. Again, the first application came about through an accident. Superconductors, showing an abrupt drop in resistivity, are unsuitable as electrical thermometers but in 1930 workers in Leiden reported measurements on a phosphor-bronze wire which exhibited a gradual drop in resistance from 7 K downward to the lowest temperatures. We were just then engaged in calorimetric measurements in the helium range and such a resistance thermometer of negligible heat capacity appeared as an immense boon. The first sample proved indeed to be an excellent thermometer but we were in for an unpleasant surprise when ordering further reels. None of the wires showed the desired properties. It turned out that the original wire had been drawn through a die previously contaminated with lead and that in the process traces of this superconductor had been distributed in minute threads along the wire.

Nevertheless, this misfortune had led to efforts of making superconductive metal samples with a gradual resistance change and these came in handy for other devices. We will only mention an extremely sensitive bolometer, an instrument for determining

10·1 The cryotron; a superconductive computer element which in its simplest form consists of a lead strip (the control) which, separated by an insulating layer, is crossed by a tin strip (the gate).

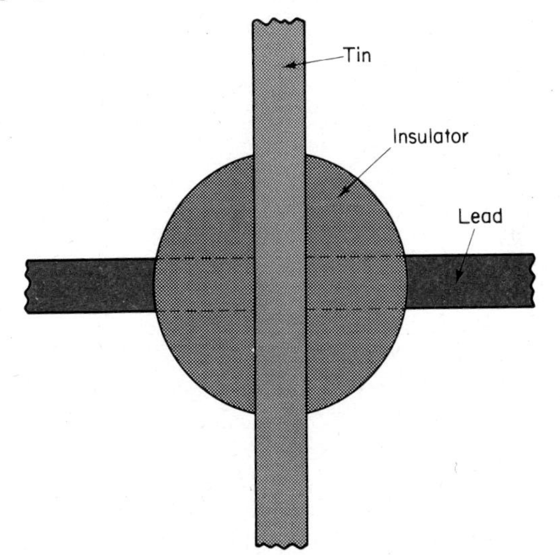

minute quantities of heat radiation which is useful in astronomic measurements and which owed its early development to efforts at providing a detector for homing missiles that will follow a heat source such as the exhaust from a jet.

All these were to some extent fringe developments, but in 1956 a device was pioneered which led to the investment of much larger sums of money than had so far been expended on superconductivity. It was a switching element which became known as the 'cryotron'. In fact, such an arrangement was not new. In 1935 Mrs. Casimir-Jonker had made use of a little lead coil wound around a tin wire to quench magnetically the supercurrent in the tin circuit. Since tin has a lower critical value than lead, the magnetic field created by a current in the lead wire did not destroy its own superconductivity. More than twenty years later D. A. Buck at a government laboratory in America proposed to use this device as switching element in a digital computer. In its simplest form it consists of a strip of tin, acting as the 'gate' the opening and closing of which is 'controlled' by a lead strip laid across a thin sheet of insulating material being interposed between the two (figure 10·1).

Modern evaporating techniques allow the strips to be made extremely thin, which means that the response time is very short. Moreover the tin gate can be made part of a superconductive ring in which a persistent current runs. By judicious arrangement of the circuit, these rings with their controlled persistent currents form an ideal 'memory', and the whole system, a rapidly operating digital computer with a vast memory from which information can be extracted with great speed, fulfils the dream of any computer engineer. In addition liquid helium had by 1956 become a commercial and cheap commodity.

Thus the scene was set for large-scale technical development and industry was not slow in taking it up. In fact, one of the biggest computer firms set to work so energetically that its competitors were discouraged from entering the unequal struggle which was further complicated by the fact that the U.S. government had applied for the key patent. After spending something approaching two million dollars the big firm closed down this computer project and the cryotron was dead. In their eagerness to go ahead, the company had assembled a team of excellent scientists who had been employed in other activities, overlooking however the minor detail that none of them had any long experience in superconductivity. Unlike conventional projects in electrical engineering, superconductivity is somewhat 'exotic' and its management requires men who have lived with it and who are familiar with its peculiarities.

One of the chief difficulties lay in evaporating trays of switching elements of identical characteristic and the natural tendency was to try to produce very pure tin gates. And here came the snag. The more the tin was refined, the less homogeneous did the gates become. This was a feature well known in previous researches on superconductivity. A metal can be made progressively purer but in doing so the purity of the individual batches tends to vary more and more. On the other hand, it is relatively simple to make metals homogeneously impure. In our own, purely academic, work we had begun to evaporate alloys for the gate and after a good deal of trying got them of sufficiently equal impurity to offer reasonable prospects of further development. The use of alloys offers an additional advantage in having in the normal state of the gate a high resistance which can be conveniently matched to an impedance at room temperature. On the other hand, making tin gates purer and purer led to a progressively lower resistance in its normal state which makes it increasingly difficult to sense its opening and closing.

However, this development came too late. The industrial project had been closed down and other new computers had been developed. We have spent some time in relating the cryotron saga because it provides a valuable lesson. With the new results and with competent thinking the cryotron project could have come to a successful conclusion. But it is impossible to persuade any board of directors to pick up a project on which their largest competitors have burnt their fingers to the tune of over a million dollars.

The latest successful development in applying low current superconductivity has the advantage of not involving large sums of money – as yet. A student at Cambridge, Brian Josephson, while still an undergraduate, had published a note in *Nature* that laid the foundation of his work for which he received the Nobel Prize at the age of 23. Josephson's achievement was based on the quantisation of a supercurrent which we dealt with in the last chapter. He combined this phenomenon with another quantum effect called 'tunnelling' which is the quantum statistical probability of a particle passing through a potential barrier, first observed in nuclear reactions. Radioactivity, in fact, represents the quantum-mechanical tunnelling of nuclear particles that have pierced the barrier in this manner. Metal electrons, too, can tunnel by passing through a very thin insulating barrier into an adjacent piece of metal. Tunnelling from one superconductor into another or into a normal metal also occurs, and it has provided direct evidence for the energy gap mentioned earlier and for the quantisation of supercurrents. In chapter 7 of this book the wave description of a particle was used and we can say that the electron wave leaks through the insulating

10·2 The SQUID (Superconductive QUantum Interference Device) is characterised by tunnel junctions which, while remaining superconductive, allow magnetic flux quanta to pass.

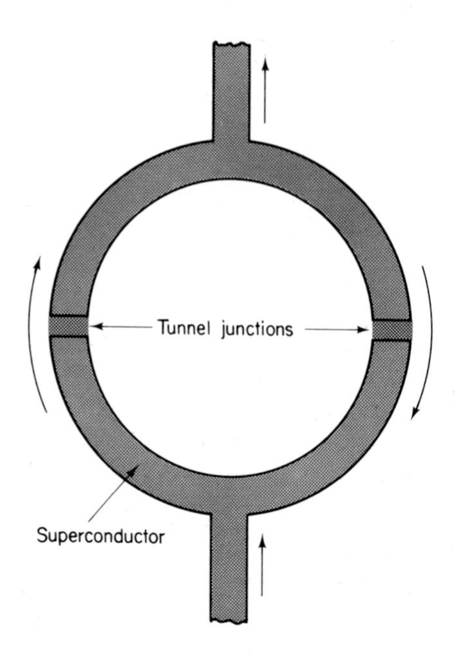

layer. Electron pairs, too, can tunnel but in addition account has to be taken of the long-range coherence of these waves. Tunnel junctions between superconductors have the unique property that they can be superconductive while also allowing magnetic flux to pass. This Josephson effect thus allows the counting of individual flux quanta by a macroscopic device.

We cannot go here into the details of this mechanism which essentially uses the quantum properties of electron waves on a large scale, as for instance their interference which corresponds to the optical interference of light passing through two slits. An arrangement based on this principle is called a SQUID, Superconductive QUantum Interference Device. Being able to count single quanta, it allows the measurement of magnetic flux with the quite unprecedented sensitivity of about 10^{-15} tesla cm^2. Voltages

10·3 Various forms of tunnel junctions.
(a) Contact between oxidised strips of a superconductor. (b) Point contact. (c) Niobium wire with a blob of soft solder.

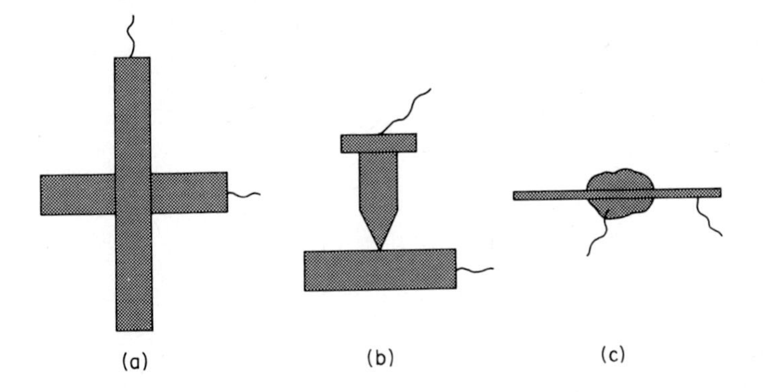

can be determined to an accuracy approaching the statistical thermal fluctuations and the same is true for currents. It is clear that there will be many applications for this quantum-mechanical macroscopic instrument, only a small number of which have so far been explored. Apart from the necessary use of liquid helium, the squid is a very simple instrument employing in its basic form a superconductive ring with two tunnelling junctions and two leads (figure 10·2). For some purposes only one such 'weak junction' is required and this is the only part of the instrument demanding some ingenuity. Various forms of this junction have been used (figure 10·3) demanding occasionally a bit of experimentation or careful adjustment. However, these are minor matters which are unlikely to detract from ever wider use of this new and sensitive instrument.

However important these weak current devices may turn out to be, the real prize for employing superconductivity in electrical engineering lies, as Kamerlingh Onnes had foreseen, in the creation of high magnetic fields, and that means strong resistance-free electric currents. In the previous chapter we mentioned the abortive attempts to achieve it, and hopes of solving this problem were gradually given up in the late thirties—until they were suddenly revived by a chance discovery twenty years later.

In 1955 a young physicist, G. B. Yntema, at the University of Illinois wanted to make some experiments requiring a magnetic field at low temperatures and decided to use a conventional electromagnet with windings of niobium wire. Pure niobium, as was known from experiments carried out in 1937, has a critical value of about 0·25 tesla and this should have given him a suitably high field between the pole pieces of his magnet. This field actually turned out to be about 0·7 tesla but the critical value of his wire was 0·5 tesla, twice that to be expected. It seems from the published short note that Yntema was not really worried by this discrepancy but rather emphasised the success of his magnet, recommending it for similar uses. Nothing happened for another five years and we may take it that this profound break-through in superconductive engineering was not recognised by Yntema or anyone else. In any case the critical value was just twice the previously published one, and that was from a measurement 18 years earlier.

Before we can deal with the true significance of Yntema's observation and its bearing on subsequent events, we must return to the alloy story. A great deal of work had been done in this subject in the thirties and in particular Shubnikov at Kharkov and the Oxford laboratory pursued these researches vigorously. By then personal communication with Russia had become practically non-existent and we had to compare results simply from published data. These were much alike, with gradual penetration of magnetic flux, except that Shubnikov's latest results came nearer to the pattern later postulated by Abrikosov. The Kharkov team evidently had excellent metallurgists to aid them but all this came to an end when Shubnikov was arrested on trumped-up political charges and disappeared into Stalin's prisons from which he never returned. His widow, Olga Trapeznikova, who also had been his colleague, survived the purge and was present at our first meeting with the Soviet colleagues in 1957 when a toast was drunk to the memory of her distinguished husband.

Years earlier at that memorable meeting of the Royal Society in 1935, we reported the curious fact that, unlike the behaviour of pure metals, magnetic flux that had penetrated alloys was not expelled on reducing the external field but remained trapped in the metal. This suggested that our alloys were not magnetically homogeneous but contained a network of threads which had a critical value much higher than the bulk of the material. As the external field is reduced, persistent currents are set up in the loops of the network and flux is trapped like water in the meshes of a 'sponge'. This model seems to have stood the test of time although two decades later the issue was unnecessarily complicated by its rediscovery under the name of 'flux-trapping', which is essentially identical with the old 'sponge'. What really matters is that the presence of filaments with high critical value in certain materials forms the basis of the new technology of superconductive power engineering.

When we use instead of the term 'alloys' the words 'certain materials', it is to take into account the fact, also discovered during our early work, that traces of impurity or even mere physical strain in pure metals result in the formation of a 'sponge'. In fact, much time and effort were spent in getting rid of inhomogeneities, particularly in mechanically hard metals such as tantalum and niobium. The reason for these efforts was to determine from magnetic data the thermodynamic parameters for which samples with an undisturbed crystal lattice were required, and as we pointed out earlier even solid solution alloys can be made homogeneous enough to show no sponge-like behaviour.

With the discovery of type II superconductivity and that of high current-carrying materials following shortly upon each other, a good deal of confusion arose which still, to some extent, persists. The basis of both phenomena is the penetration of flux into materials with negative surface tension. Unfortunately the experimental realisation of pure type II was difficult and long delayed even after its existence had been predicted by Abrikosov. When at last we obtained a perfect specimen it exhibited all the properties postulated by the theory (figure 9·7) and in particular it showed

10·4 Properties of heavily strained superconductors. (a) The magnetisation curve shows strong hysteresis, with pure type II (broken black line) shown for comparison. (b) The sponge structure ensures much increased critical currents up to very high fields.

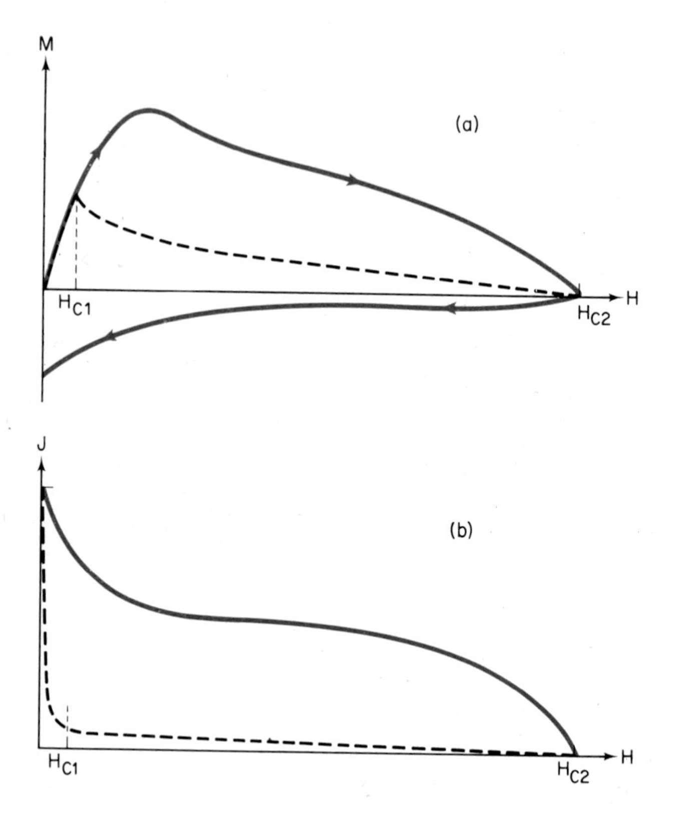

the two well defined critical fields. The most important feature for the present discussion is the complete magnetic reversibility. On the other hand, the magnetisation curve for the same material in the strained state was well known. It shows instead of the sharp lower critical field a rounded peak and a more gradual approach to the upper critical value which, incidentally, is the same as for the pure sample (figure 10·4). Even more striking is the pronounced irreversibility on the return to zero external field, all these features being characteristic for its sponge structure. However, most important is a comparison of the critical currents (figure 10·4 b). That of pure type II drops to quite small values because, on applying a current, the Lorentz force sweeps the magnetic flux out of the sample and thus renders it useless as a current carrier. In the strained sample, on the other hand, the sponge traps the flux lines and the critical current remains high to very high fields. The distinctive 'shoulder' is a significant indicator for a potentially good industrial material.

This provides the explanation of Yntemas's accidental discovery. His niobium wire was heavily strained and contained sufficient filaments of high critical value to provide a strong enough supercurrent for his magnet. The significance of Yntema's observation, however, passed unnoticed until, in 1960, S. H. Autler at M.I.T. published a full paper on the use of niobium for superconductive magnets, mentioning a solenoid yielding up to 0·43 tesla and an iron magnet with niobium windings producing a field of 1·4 tesla in the pole gap. He did not seem to be aware of Yntema's work five years earlier but fully recognised the importance of strain, predicting that strained niobium wire might retain its superconductivity up to 0·8 tesla.

This, at last, was the real break-through into the field of high-field superconductive technology. In the following year J. E. Kunzler of the Bell Telephone Laboratories gave an extensive report of experiments with different substances and their suitability for high field application. He then went on enumerating possible applications such as the use of powerful magnets for thermonuclear fusion and superconductive power transmission, pointing out that the measurements as well as the new way of thinking were only a few months old and that shortly many more applications would come to mind. This shows how emphasis of superconductivity research had suddenly changed from investigation of basic properties to technological problems, foreshadowing large-scale expensive projects. The question has often been asked whether, if the Oxford and Kharkov experiments had not been interrupted by the war, we would have had powerful superconductive magnets a quarter of a century earlier. I think the answer to this is definitely 'No'. At that time there existed neither the essential cryogenic background with unlimited amounts of cheap liquid helium nor the technological demand. Anyone suggesting then the expenditure of tens of thousands of pounds on such a development would have been considered mildly insane. It needed the war, jet propulsion, atomic bombs and space exploration to accustom the taxpayer to the idea of costly research.

With Kunzler's review article the search for technologically promising substances was on. When in 1963 a conference was held at Colgate University in America no less than 25 papers on high-field superconductivity were read by authors from all over the world and from then on work in this field has escalated at a truly amazing rate. Less than ten years later more than one thousand superconductive alloys and compounds had been tested and the metallurgy of many of these had been investigated in detail. This wealth of information posed the serious problem of judicious selection. There is a long and expensive way from knowledge of suitable superconductors to the production of miles of technological advantageous and reliable wire or cable for industrial use. High critical values, the price of the basic materials, mechanical properties and ease of manufacture have all to be considered before the right choice can be made. This choice carries an enormous financial responsibility because, far in excess of the already expensive pioneering development, fabrication on an industrial scale involves very large sums of money, usually requiring government contracts. In less than a decade superconductive engineering had become very big business indeed.

Of the large choice of possible materials only three have reached the manufacturing stage; they are alloys of niobium with zirconium or with titanium and the compound Nb_3Sn, with the second being now much in use but the last being the likely winner. The first two alloys can carry currents up to 10^5 amperes per cm^2 but giving out at 10 tesla or less while Nb_3Sn can carry twice this current up to 15 tesla or more. The first two materials owe their temporary success to comparative ease of manufacture whereas Nb_3Sn has posed formidable fabrication difficulties. It is extremely brittle and cannot be wound into coils straight away. Coils therefore have first to be wound using the component materials and then have to be reacted *in situ* at an elevated temperature. In spite of this drawback, this process is now generally adopted with niobium and bronze being used in the primary step of making the coil which is then heated to about 700°C. At this temperature, tin from the bronze diffuses into the niobium metal forming the compound Nb_3Sn.

A coil made of the superconductive metal by itself presents a serious risk when for any reason the critical value is exceeded locally. Then a strong current is quenched suddenly and the large magnetic energy stored in the coil is released in the form of heat which has been known to not only evaporate the liquid helium surrounding it but actually to melt the coil. The only possible safeguard is to provide an efficient shunt of very pure copper which can take up the current surge until the interrupted section is 'healed' by field re-adjustment and cooling. All industrial materials are therefore constructed in the form of 'composites' in which the superconductive wire is embedded in a matrix of copper. Since supercurrents tend to run on the surface, the superconductor itself is split up into a multitude of fine wires. It should be noted that, since the resistance of these filaments is strictly zero, any metal with finite resistivity, however small, acts as a perfect insulator. Thus no other insulation than the copper matrix is required, which simplifies not only the winding of the coil but also its heat treatment.

When alternating magnetic fields are required, the changes in direction of magnetic flux are bound to quench temporarily the superconductive state leading to strong eddy currents in the copper matrix. These undesirable effects can be minimised by twisting the superconducting filaments into a spiral. Thus a modern supercon-

10·5 Industrial superconductive compound materials. (a) Twisted filaments. (b) 40,000 filament compound material is drawn to 0·2 cm outer diameter. Etching away the bronze reveals the individual filaments. (c)

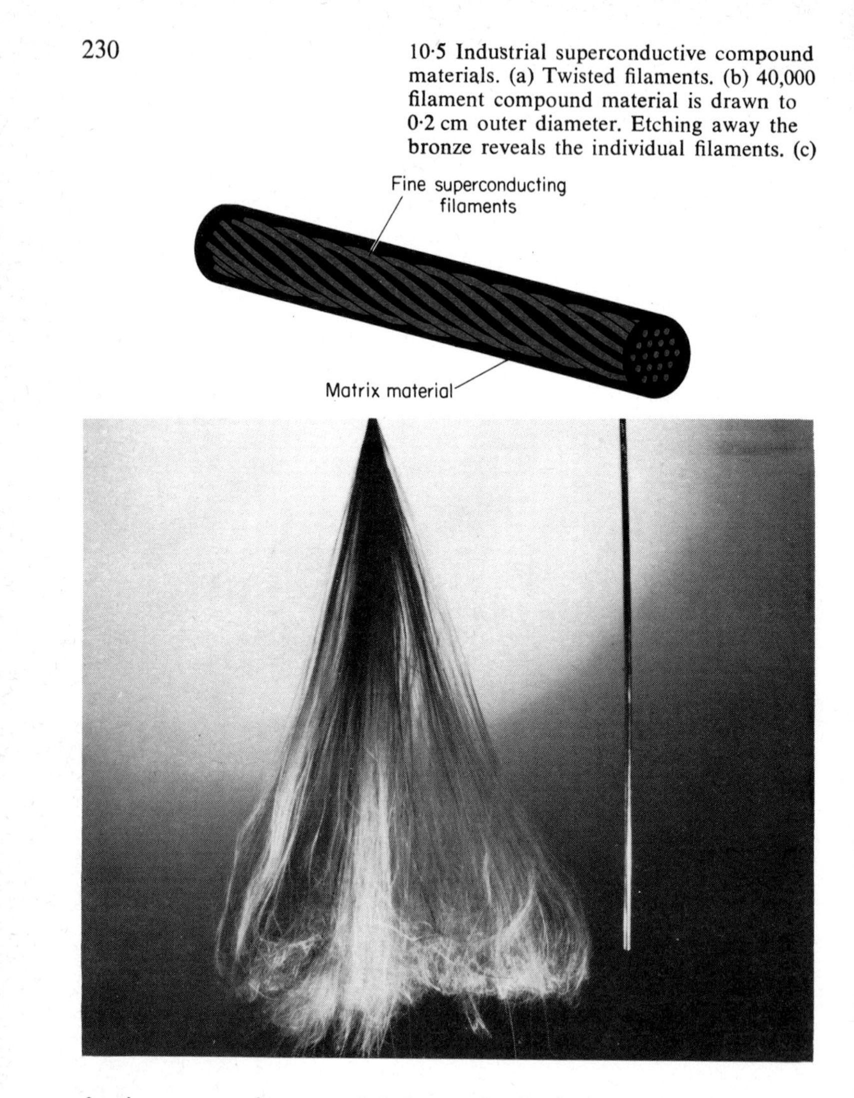

ductive composite material for technological use consists of a copper cable into which strands of tens of thousands of twisted filaments are imbedded. Since the bronze from which the tin has been abstracted for making Nb_3Sn filaments does not yield very pure copper of high electrical conductivity, additional strands of

Highly magnified cross-section of the same wire showing the bundles of filaments and strands of high conductivity copper.
(d) The same material rolled into ribbon shape.

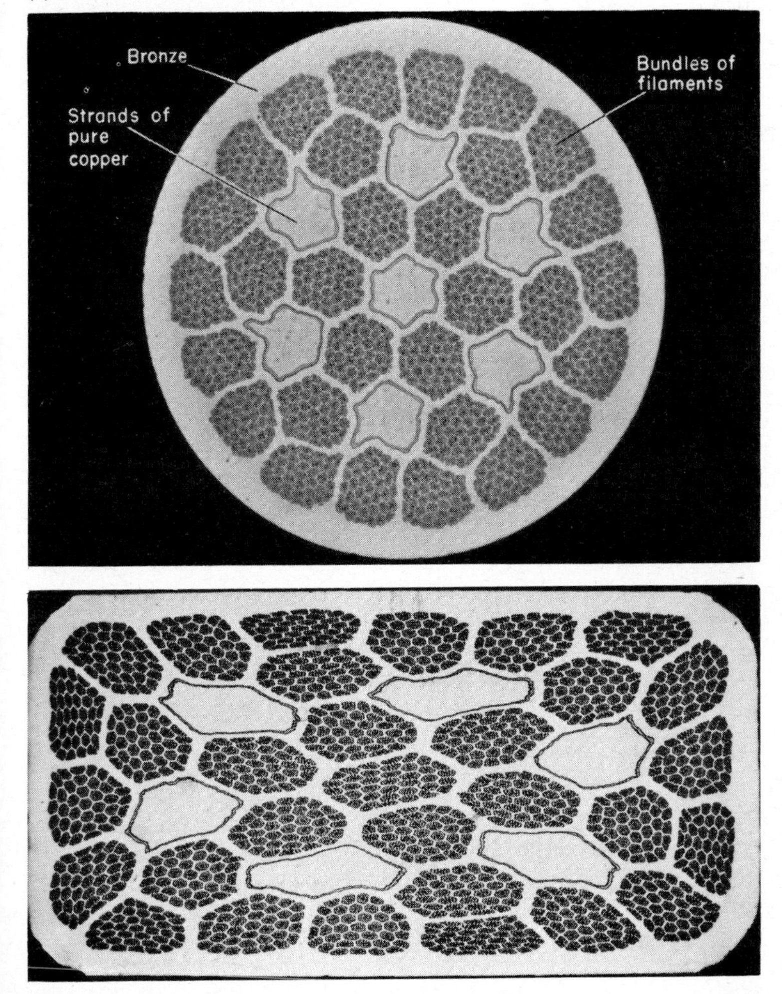

pure copper have to be distributed throughout the compound material. Finally, the finished composite may have to be shaped into rectangular cross-sections in order to achieve a good packing factor within the coil (figure 10·5).

10·6 The superconductive revolution. Conventional water-cooled copper solenoid for adiabatic demagnetisation. At a power consumption of 1,000 kilowatt it produced 5 tesla. It can now be replaced (inset) by a small superconductive solenoid of 5 cm inner diameter, delivering 10 tesla.

In spite of the very considerable fabrication difficulties outlined above, materials complying with these stringent specifications are now being manufactured on an industrial scale. Their development was the necessary prerequisite for embarking on the very ambitious and extensive technological projects to which we must now turn. All these large-scale applications are based on the use of high magnetic fields which with the help of superconductors can be maintained without loss. The early uses were confined to experiments where the investigation itself was carried out at low temperatures such as in demagnetisation cooling of either electron or nuclear spins. Later, and this is the far more important development, cryogenics has merely provided an adjunct, though an essential one, for cooling the superconductive coils to furnish a high field that can be turned to a great number of uses, none of them necessarily connected with low temperature problems.

It would, however, be erroneous to assume that in these last cases superconductivity is nothing better than a substitute for conventional electric engineering. As we shall see, the use of superconductors allows applications which never could have been achieved with normal metals. Taking first the case of adiabatic demagnetisation experiments. Here conventional water-cooled solenoids have been stretched to the limit of performance both as regards the current supply and particularly in carrying away Joule heating from the copper coils. Using a superconductive solenoid instead, a modest power supply is sufficient for energising the coil and this can be short-circuited after a persistent current of the required strength has been established. Water cooling is not required and, since the whole apparatus is immersed in liquid helium in any case, the superconductive solenoid can be small since it is located immediately around the cooling substance. Comparison of this device with the conventional method in size, energy consumption and simplicity of construction is most impressive (figure 10·6).

The next stage in the technological use of superconductors were the magnetic fields employed in nuclear research, at first in bubble

OXFORD INSTRUMENTS
OXFORD ENGLAND

chambers and then in storage rings. In the first case they provide the very strong fields needed to bend the tracks of fast particles while in the second case particles are confined to flight in circular evacuated tunnels of close to one kilometre diameter. The object of this arrangement is to accelerate the particles by periodic bursts, the acceleration as well as the confinement to the circular path being produced by magnetic fields. Since the large diameter of the rings necessitates enormous field energies, this is a project far beyond the capability of conventional magnets and for which superconductive coils will be the only answer. Into the same category fall 'magnetic bottles' and similar devices in which a hot 'plasma', a gas of highly energetic nuclear particles, can be held. These contraptions whose magnetic walls are created by high fields are the forerunners of machines to create energy by nuclear fusion, a controlled reaction using the power of the hydrogen bomb. Closely related to these projects is the problem of superconductive energy storage. When discussing the hazards of electric breakdown in a superconductive coil we have referred to the large amount of energy inductively stored in a persistent current in such a coil. The conventional means of storing electricity in lead batteries has many disadvantages; it is unsuitable for storing vast quantities and is limited in the rate of release. Apart from the cryogenic supply, there is no limit for electricity reservoirs represented by superconductive coils from which moreover extremely strong currents can be drawn in very short pulses. Contemplated projects envisage huge underground coils of as much as 100 m diameter and storage capacities of 30 million kilowatt-hours.

Not only power storage but also power transmission, using superconductive cables, is being considered. Since the whole length of the line has to be refrigerated, short distances requiring very high energy flow, such as highly concentrated industrial areas, are obvious areas of application. Limited lengths of prototype cables have already been operated successfully by various firms (figure 10·7). Another field of industrial importance is the separation of weakly magnetised matter as in kaolin purification or sewage treat-

10·7 Superconductive transmission cable with helium cooling and liquid nitrogen shielding.

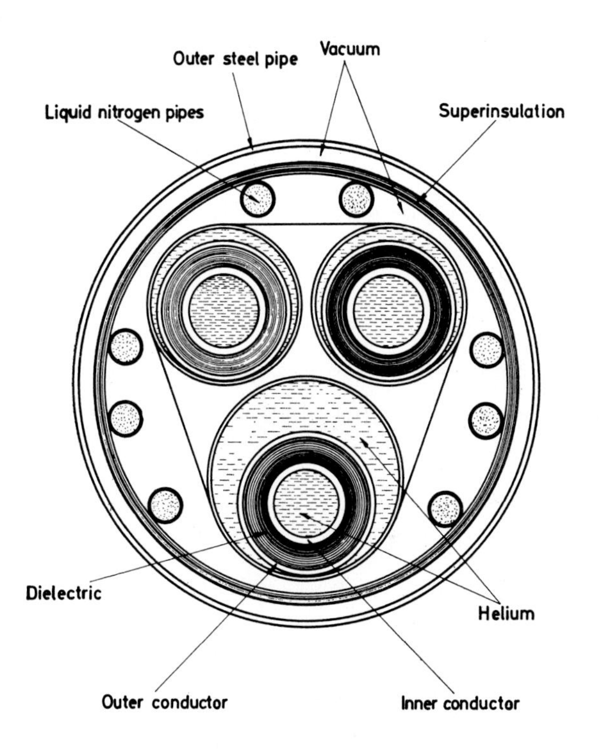

ment. Here large quantities of fast-flowing material have to be processed in strong magnetic fields, which to maintain by conventional means is prohibitively expensive. Strong persistent currents, on the other hand, can be kept running without loss since the power required for the actual separation process is small.

It was, of course, inevitable that, with suitable materials becoming available, the question of superconductive power machinery would arise. Simple calculation shows that the provision of liquid helium would not be a serious drawback when the saving in resistive heating and weight of equipment is considered. The

10·8 The Faraday disc and a 3,250 horse-power homopolar d.c. motor under construction. The disc is just being mounted in the ring-shaped container which holds the super-conductive field coil in a bath of liquid helium.

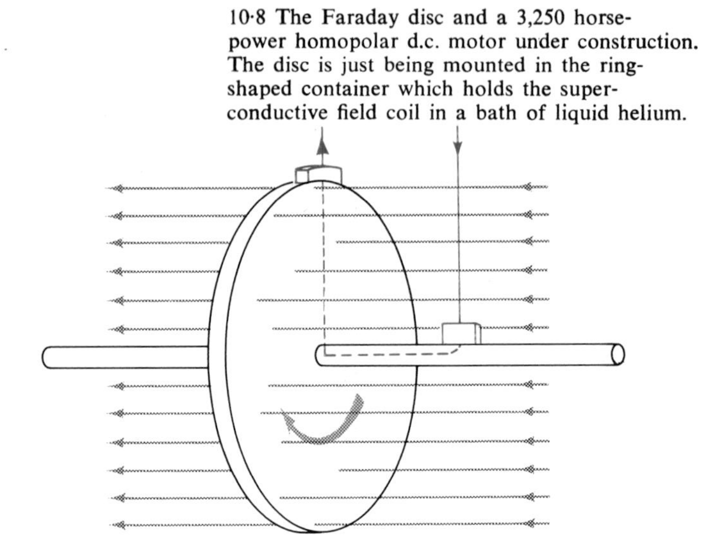

difficulty lies in the design of reliable rotating units for motors and generators. This problem has partly been solved and superconductive motors of over two megawatts have been successfully operated without trouble over long periods. As was to be expected, the design of these machines differs radically from the conventional type. The machines work on the homopolar principle, well known as 'Faraday disc'. When a circular copper plate is made to rotate around its axis in a perpendicular magnetic field, a current flows from the axis to the periphery of the plate where it can be drawn off (figure 10·8). The arrangement thus represents a dynamo and, when reversed, will act as an electro-motor. It has the great advantage that no rotating field is required and that the magnetic field threading the disc can be produced by a stationary superconductive coil within a vacuum vessel. The first motor of this kind was commissioned by the British Admiralty and work began early in 1965 with the prototype model of 2 hp (about 1500 watts) running 18 months later. It was soon scaled up to 50 hp (37 kW) and then followed a motor of 3,250 hp (nearly 2·5 MW). The obvious application of such machines is in marine propulsion where a prime mover will drive a completely superconductive generator-made unit as has recently been operated successfully.

Meanwhile, development has not stood still and in America a prototype generator with a helium-filled superconductive rotor has been successfully operated. The consensus of opinion among power engineers engaged in these projects is that superconductive machines have turned out remarkably troublefree and that there is no reason to doubt that they can successfully be scaled up for use in large power stations. In fact, with increasing power concentration due to the energy input from nuclear reactors, the superconductive generator may soon become the only type of machine to manage this large energy flow adequately. It is estimated that above a rating of about one and a half million kilowatt the superconductive generator will become more economical than the conventional machines.

All these large scale applications of superconductivity, however unusual, have some link with earlier technologies. This cannot be said of the most recent one: fast ground transportation, made possible by magnetic levitation. Levitation by making use of diamagnetic repulsion was demonstrated in the laboratory many years ago (figures 9·2 and 9·3) but suggestions for its use had been limited to gyroscopes and low-temperature frictionless bearings. Things changed when it was realised that the base for the repulsive force need not be superconductive in itself but that the magnetic image could be effective in a normally conducting base, provided that the superconductive magnet moved over it at speed. In the last few years models and full-size prototype vehicles have been produced. We cannot here go into the intricacies of design of either the vehicle or method of propulsion which is achieved by a travelling magnetic field, called a linear motor. The vehicles start off on wheels until their speed gets high enough for the superconductive field coils to lift them off the ground when they become magnetically floating (figure 10·9).

It has always been felt that for large-scale travel between populous urban centres a few hundred miles apart ground transportation would be the ideal solution if only the rate of travel could be increased by frequent departures and, above all, higher

10·9 Test vehicle kept magnetically floating
by superconductive levitation coils.

speed. The two centres of Tokyo–Yokohama and Kyoto–Osaka, each with well over ten million inhabitants and 350 miles apart, are a typical example and it is not surprising that Japan has taken the lead in fast trains. Their crack train, the *Shinkanzen,* runs at 20-minute intervals, covers the distance in a little over three hours and has been transporting over half a million passengers a day. However, the Japanese engineers feel that with this frequency of travel at 130 miles per hour they are close to the safety limit although the trains are already completely computerised. After examining various systems they have decided that the magnetically levitated vehicle, using superconductive repulsion will provide the ideal solution. The chief advantage of a train, safe guidance, will be maintained while any direct contact with the ground is eliminated (figure 10·10).

Concluding our account of large scale application of superconductivity, we are suddenly faced with a vast new technological potential based on temperatures close to the absolute zero which ten years ago, when the first edition of this book was published, could not even be dreamed of. Today we stand at the threshold

10·10 Prototype study for a superconductively levitated high-speed train.

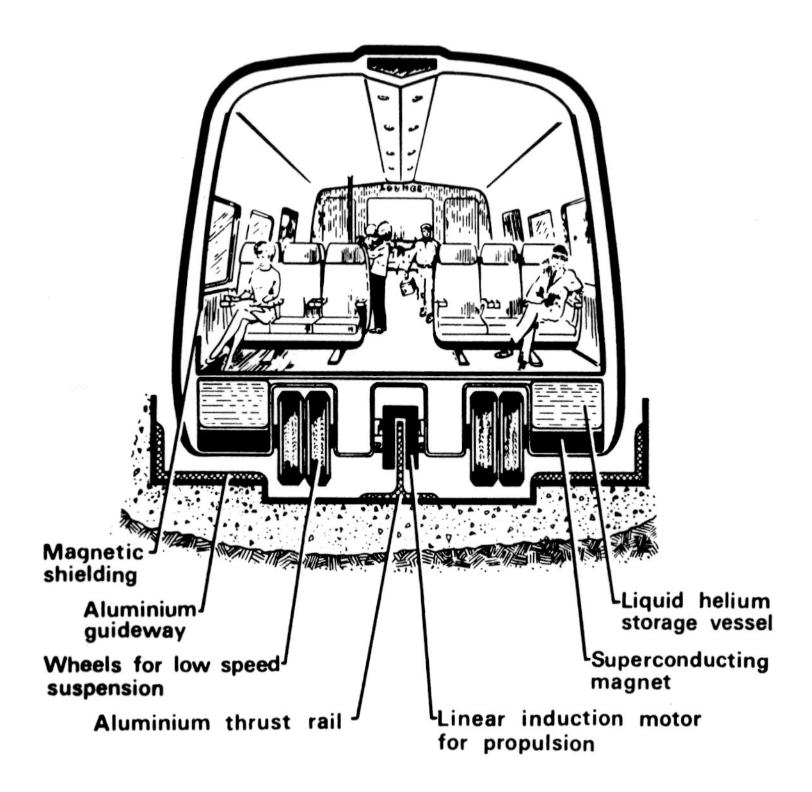

of an immense break-through which is no longer an engineering one. The essential developmental problems have indeed been solved and the prototypes tested. The next step rests with the treasuries of governments and their preparedness to spend vast sums of money. Superconductivity has ceased to be a scientific problem and has now become one of political handling of economic issues.

11 Superfluidity

In retrospect it must appear quite inconceivable that thirty years should have passed between the first liquefaction of helium and the discovery of its most spectacular property, superfluidity. In those thirty years thousands of experiments with liquid helium were made in laboratories all over the world and in many of them the experimenters must have watched a dramatic change in the aspects of the liquid taking place as it was cooled. There is every reason to believe that Kamerlingh Onnes saw it on that memorable 10th July 1908. When almost a quarter of a century later McLennan, Smith and Wilhelm in Toronto finally mentioned it in print, they still failed to draw the obvious conclusion. Describing the boiling of liquid helium under reduced pressure, as its temperature fell through the region of the density maximum discovered by Onnes, they said: '. . . the appearance of the liquid underwent a marked change, and the rapid ebullition ceased instantly. The liquid became very quiet. . . .' (figure 11·1). Again, as in the case of superconductivity, it was the enormity of the discovery which prevented it being made. The correct inference could have been drawn by a first year physics student, but which mature and experienced physicist dared seriously to suggest that the heat conduction of the liquid had suddenly increased a millionfold. Nevertheless that is exactly what had happened.

Let us go back for a moment to the experiments of Dana and Onnes in 1924 which we mentioned in chapter 4. Just below the temperature of the density maximum at 2·2 K they had found values of the specific heat which were so high that they did not dare to publish them. They felt sure that something had gone wrong with their measuring equipment. Dana had to return to America and shortly afterwards Onnes died. Another six years went by before the question of the specific heat of liquid helium was settled by his successor Willem Hendrik Keesom and a German guest worker Klaus Clusius. In Communication No. 219e from the Leiden Laboratory they published their results which showed a huge anomaly in the specific heat. Because of the similarity of the curve to the shape of the Greek letter λ, it has become

11·1 Liquid helium (*left*) above and (*right*) below, the lambda-point. Boiling ceases and superfluid helium runs out through the fine pores in the bottom of the vessel suspended above the helium bath.

known as the lambda-point. This peak in the specific heat occurs at exactly the same temperature as the maximum of the density (figure 11·2).

Keesom realised immediately that this must signify some fundamental change in the nature of the substance and, understandably enough, many people suspected that helium had, after all, a triple point and that the phase below 2·2 K might, in fact, be crystalline. Admittedly, it was mobile but instances are known where the crystal planes are so smooth that they are continually slipping

11·2 At the lambda-point the density and the specific heat of liquid helium show maxima.

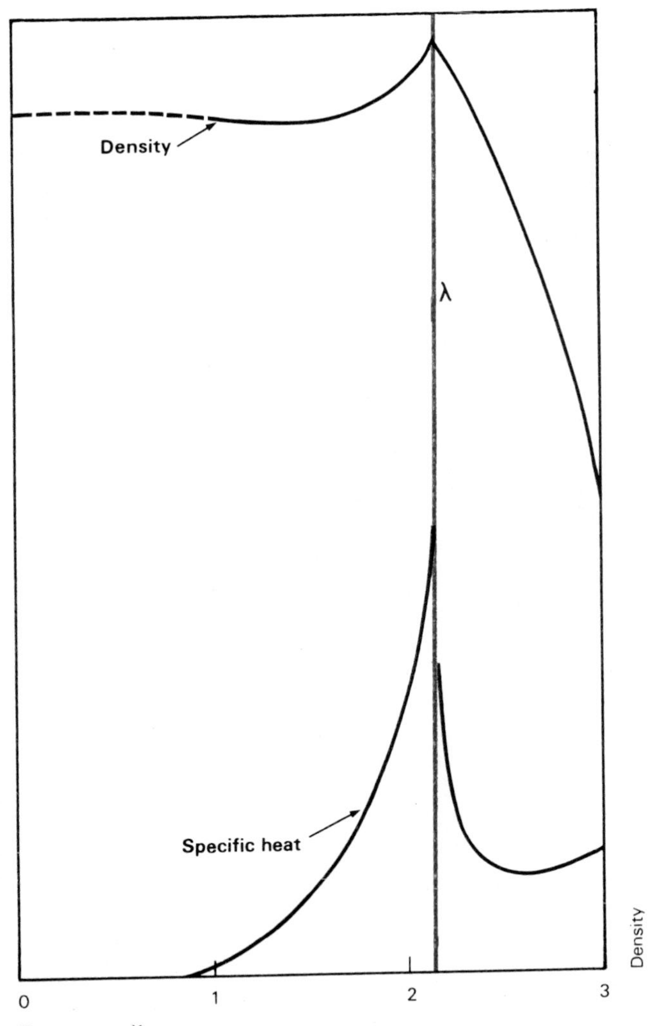

and provide an aspect of mobility. Liquid helium, they argued, owing to its low boiling point, must be ideally pure and it might be the extreme case of such a 'liquid crystal'. This explanation, elegant and intriguing as it was, turned out to be wrong. The regular pattern of a crystal lattice must show itself in a regular pattern of diffracted x-rays, but when a few years later helium below the lambda-point was subjected to x-rays, by Taconis in Leiden, they revealed no such pattern. It had to be accepted that below as well as above this temperature the substance is a liquid, but a liquid which undergoes some profound change at 2·2 K. What this change was Keesom could not say. All he could do was to make a distinction between the two forms of the liquid, that above and that below the lambda-point, by calling the former He I and the latter He II, a notation which has been retained and which we shall use in the following. It also means that the diagram of state of helium is unlike that of any other substance. There is no triple point because the melting curve ceases to vary with temperature at 25 atmospheres of pressure (figure 11·3). Consequently the latent heat of fusion disappears and, as absolute zero is approached, melting becomes a purely mechanical process which can only be brought about by pressure changes. The liquid phase is extended to zero temperature and it is divided into the two domains of He I and He II.

It was with reference to these two liquid forms that the remarks of the Toronto team, quoted above, were made and under these circumstances it must appear even more extraordinary that neither they nor anyone else should have recognised the significance of the observation. To make things worse, three years later Keesom not only came into the possession of the strongest possible clue for the solution of the Toronto mystery and even published it but, for the time being, he and everyone else passed it over.

In collaboration with his daughter Anna Petronella he had repeated the specific heat measurements with greater accuracy. On passing through the lambda-point they noticed that the character of their data underwent a sudden change which they correctly

11·3 The diagram of state of helium showing the lambda-line which separates He I and He II.

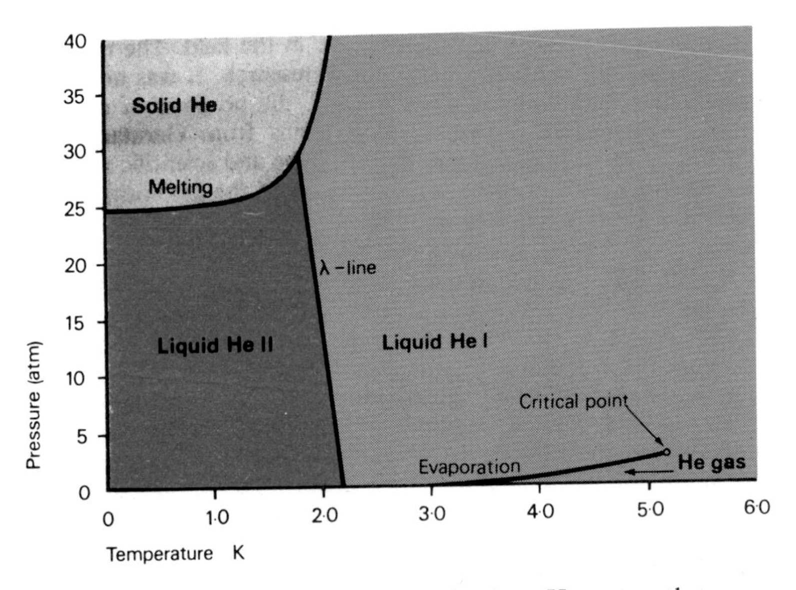

interpreted as a rise in thermal conduction. However, the connection with the remarks of McLennan and his colleagues was not recognised. The next step of the Keesoms was a proper determination of the heat conductivity of He II for which they constructed a conventional type of apparatus suitable for a poor conductor. When this apparatus failed, it was finally realised that the thermal conduction of He II must be very large.

It is significant that the idea of an enormous heat conductivity of liquid helium only occurred when it had become quite inescapable. That a dielectric liquid should suddenly become a better conductor of heat than copper or silver could simply not be comprehended on the basis of the known physical phenomena. Once this fact had to be accepted, it became clear that the door had been opened to new discoveries. A stage had been reached similar to the first indication of superconductivity, except that now a number of experienced low temperature laboratories went into the attack where formerly Leiden had been alone in the field. The next few years brought a veritable avalanche of research. It was now 1936 and as the new discoveries were made, the progress of research was punctuated by disturbing news items from Germany. The shadow of war began to loom over Europe and scientific research was becoming a race against time. Most of the astonishing properties of He II were discovered in the span of less than two years and a useful model, able to account for all these phenomena, was established. Practically all these results, experimental and theoretical, are recorded in that one famous volume of *Nature*, No. 141. To anyone who was not in the game at the time the telescoped evidence presented in it has the aspect of a hopeless tangle. However, nothing could be more misleading. The attack on the problem is an excellent example of scientific co-operation at its best. There was fierce competition between the different laboratories but they kept each other fully informed of the progress achieved, often over the telephone and through personal meetings. At these there were ruthlessly searching discussions into the validity of each others' methods but with much constructive criticism.

The first investigation which led to a new discovery was innocent enough. Allen, Peierls and Uddin in Cambridge had hit upon an ingeniously simple method for repeating the Leiden heat conductivity measurements with greater accuracy. The Keesoms had used a long tube filled with liquid helium and had recorded the temperature drop along this tube with electrical thermometers. The Cambridge arrangement (figure 11·4a) merely consisted of a glass bulb which communicated with the liquid helium bath through a tube at the bottom. The bulb contained an electric heater and when this was switched on, heat travelled through the helium in the tube into the bath. The temperature difference at the ends of the tube was indicated by the difference in vapour pressure of the liquid in the bulb and in the bath. It could be observed simply by the level difference, the weight of the corresponding column of liquid helium being equal to the vapour pressure difference.

Using this simple but very sensitive apparatus the Cambridge team not only confirmed the high heat transport found in Leiden but discovered an important additional fact. Unlike any ordinary conduction process, the heat flow through He II was not proportional to the temperature difference. In fact, the lower this difference was made the greater became the 'conductivity', reaching values exceeding that of He I by a million times. It was an important discovery which, if followed up at the time, might have led to investigations which were only carried out more than ten years later. The result eventually turned out to be quite correct, but very shortly after its publication the authors themselves called it in doubt. Something quite unforeseen had happened.

Extending the measurements to lower and lower temperature differences, it was noted that the gap between the menisci not only decreased but actually reversed its sign. In other words, the level inside the bulb rose slightly above that of the bath outside. Regarding this still as a difference in vapour pressure would have meant that the liquid heated inside the bulb became colder, and this was obviously nonsensical. The next step was to cut open the top of the bulb and in this way to repeat the experiment under conditions in which the vapour pressure above both levels was the same. The result was perplexing; on heating the level inside rose again (figure 11·4b). A quite new effect had been discovered which occurs only in He II and in no other liquid: when heat is supplied, a flow of liquid takes place in the direction of the heat source.

It was soon noticed that the level difference increased when the tube connecting the bulb with the bath was made narrower. Finally a large number of very fine channels was used by packing a tube with powder so that the liquid had to flow through a multitude of small interstices. Not only could a very large level difference be obtained in this way but when the top of the bulb was drawn out like a syringe, a jet of liquid helium was seen to be ejected from it (figure 11·4c). This rather spectacular display led to the name of 'fountain phenomenon', but we shall refer to it by the more appropriate term of 'thermo-mechanical effect' which indi-

11·4 The discovery of the thermo-mechanical effect (a and b) and the helium fountain (c). The mechano-caloric effect (d).

cates that we are dealing with a motion of mass, the liquid, produced by heat.

Since it was evident that the thermo-mechanical effect must have influenced the results of the original heat conductivity measurement these could not be trusted, and for a time it seemed as if the lack of proportionality between heat current and temperature difference, discovered in Cambridge, was due to such falsification. However, the next batch of heat conductivity data, this time from Leiden, showed that the Cambridge result was true after all.

The term 'thermo-mechanical' immediately suggests a heat engine and the possibility of applying to it the equations of thermodynamics. This was, in fact, done in the following year by H. London who correlated the work done in lifting up liquid helium in the 'fountain' with the heat required to do this. In the meantime, it occurred to us to see whether, like other heat engines, this one could be reversed. An apparatus was therefore made at Oxford which allowed He II to flow through a set of narrow channels from

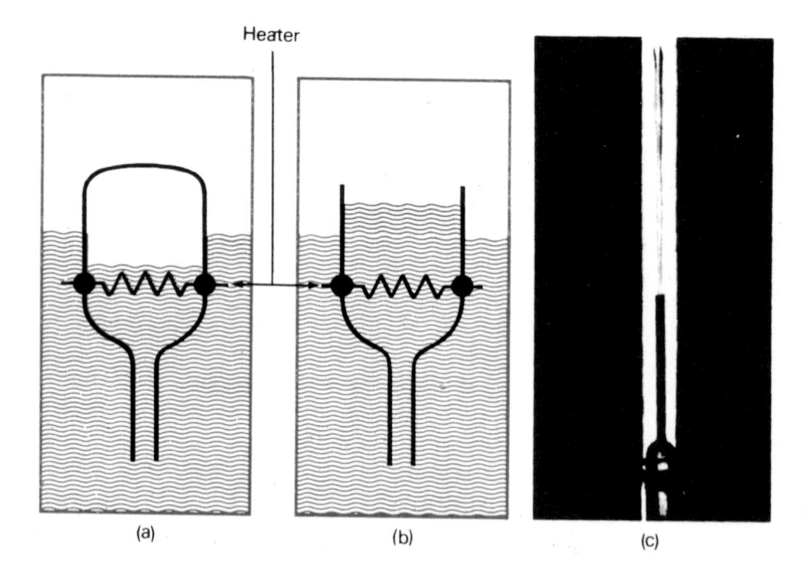

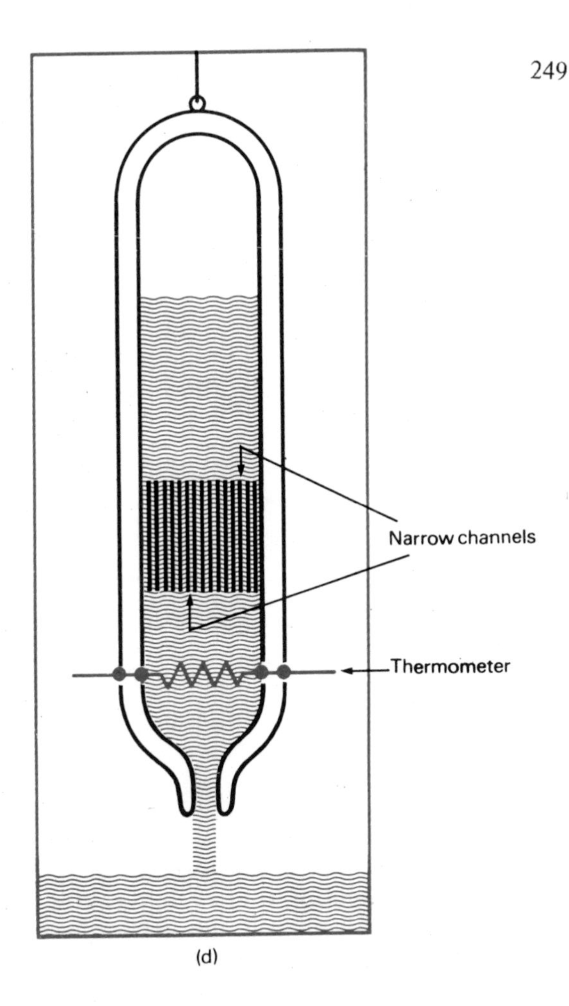

(d)

a higher to a lower level. The flow tube was provided with double walls like a Dewar vessel to prevent any stray heat influx (figure 11·4d) and a very sensitive thermometer was placed at the outflow from the narrow channels. When He II was made to flow through this tube, it was found that the outflowing liquid became cooler. This 'mechano-caloric effect' is the true reverse of the thermo-mechanical effect observed in Cambridge. Their relation is best demonstrated in a diagrammatical representation (figure 11·5) of

11·5 Diagrammatic representation of the thermo-mechanical (*left*) and the mechano-caloric (*right*) effects.

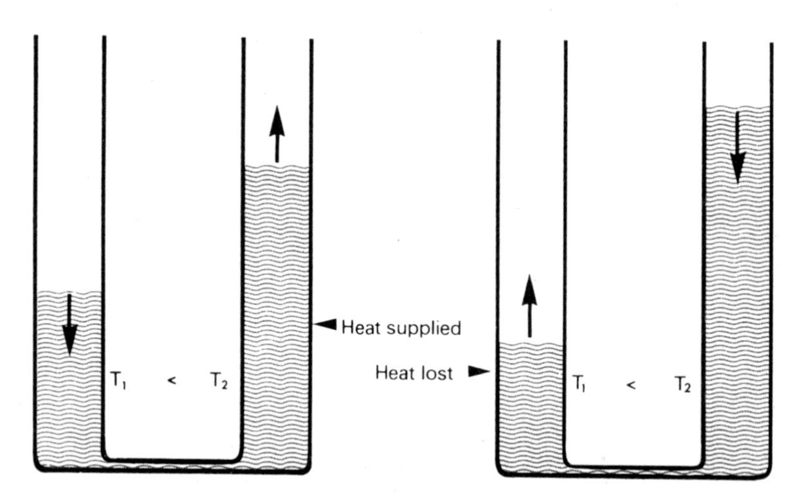

two vessels containing He II which are connected by a narrow tube. It shows immediately that there is a unique and reversible connection between the flow of mass and of heat. These take place in opposite directions.

Astonishing as these thermal effects in He II were, they did not by any means claim the full attention of those working in the field. Another remarkable event had taken place between the Cambridge heat conductivity measurements and the observation of the thermo-mechanical effect; the discovery of superfluidity.

It had been noticed in a number of laboratories that vessels containing liquid helium which were not perfectly tight, leaked very much more below than above the lambda-point. Sometimes an apparatus which appeared quite tight above 2·2 K would begin to leak badly as soon as it was cooled, and become useless. The Keesoms who were troubled by this irritating feature in their specific heat measurements in fact suggested that the viscosity of He II might be lower than that of He I. Two direct determinations of this quantity made shortly afterwards, one in Toronto and the other in Leiden which, while differing in details of their results, both showed that below the lambda-point the viscosity decreased. The drop observed was appreciable, but nothing which could stand comparison with the enormous increase in the heat conduction. In both these determinations the internal friction was measured by the damping in the movement of a cylinder or disc oscillating in the liquid.

Then early in 1938 *Nature* published side by side two articles, one by Kapitza in Moscow and the other by Allen and Misener in Cambridge. Both of these described viscosity experiments in which the friction of the liquid had been measured by flow through fine capillaries or through a narrow slit between plates. The result was the same; the liquid flow was almost non-viscous, encountering less resistance the narrower the orifice was through which it had to pass. Kapitza suggested the term 'superfluidity' for this new phenomenon (see figure 11·1).

Although superfluidity was a strange and unsuspected property of He II, it showed a distinct similarity to the high heat conductivity in that its magnitude depends on the experimental conditions. Whereas the oscillation experiments yielded a drop in the viscosity below the lambda-point to about one-tenth of its original value, in the slits it was of the order of one-millionth. Moreover, as if this bewildering variety and complexity of new effects in He II was not enough, another one was added to the list shortly afterwards.

As early as 1922 Kamerlingh Onnes observed in one of his attempts at reaching a very low temperature that the liquid levels in two concentric Dewar vessels adjusted themselves to the same height, an effect which he ascribed to distillation from one into the other. He briefly described his observation but no further notice was taken. I certainly did not suspect any connection with it ten years later when my calorimetric measurements were spoiled by a layer of helium which had evidently formed on solid surfaces. Similar difficulties were experienced in 1936 by Rollin in Oxford who, like Kamerlingh Onnes, tried to reach low temperatures by pumping off the vapour above liquid helium. He came to the con-

11·6 Film transfer in He II (a), and its application to demonstrate zero friction (b).

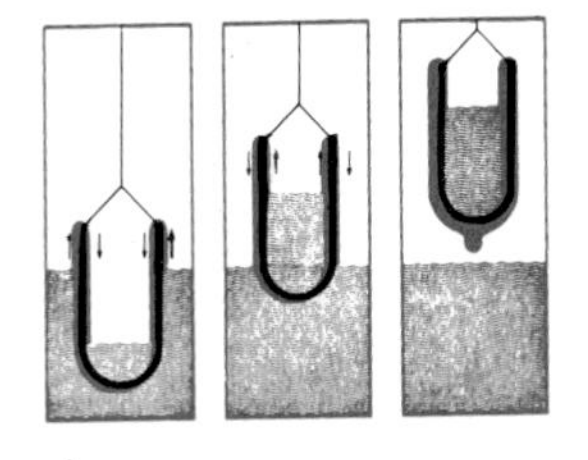

(a)

(b)

clusion that a film of helium had formed on the walls of his apparatus and, since the high heat conductivity of He II had just been discovered, he attributed the trouble with his cryostat to heat influx along the film. Two years later Lasarev in Kharkov came to the same conclusion.

At about the same time – it was now 1938 – a research student, J.G.Daunt, and I wanted to see whether Onnes's long forgotten distillation experiment could be reproduced. We filled two little glass bulbs, which were connected on top by a tube, to different levels with He II. Nothing dramatic happened and we were on the point of giving up the experiment when after more than half an hour we noticed that the lower level had risen by a few millimetres. Next we increased the connecting surface between the two reservoirs of He II by putting quite a number of fine wires into the tube; and now the transfer of liquid between the bulbs increased markedly.

Suddenly everything seemed to fall into place. There was no distillation, instead helium was being transferred by means of a surface film. Moreover, it was not the heat conductivity of the film which was high but an actual flow of liquid along it. The heat influx into the cryostats was not a conductivity effect, but due to film transfer up the walls of the vessel and subsequent evaporation at the warmer region on top. Part of this vapour then re-condensed on to the liquid carrying with it the heat of evaporation.

The next step was to demonstrate unambiguously the existence of such a mass transfer and this could easily be done by hanging a small glass beaker into the liquid (figure 11·6a). Soon the little beaker was seen to fill up with liquid until the levels inside and out were at the same height. Flow in the opposite direction took place when the beaker was raised a little, and finally, when the beaker was completely lifted out of the liquid, little drops could be seen to form at the bottom of the beaker and fall back into the bath at regular intervals. Quite apart from the astonishing spectacle itself, it was this regularity which claimed our attention, and indeed we noticed that the emptying and filling of the beaker always took place at the same rate, independent of the level difference, the length of the path or the height of the rim above the surface.

We now realised that by a lucky chance the film had provided us with the purest example of superfluidity imaginable. Measuring the film thickness, we found it only 50 to 100 atoms thick, much thinner than the narrowest slits or capillaries used so far. The characteristic feature of superflow thus was transport completely free of friction, taking place at a 'critical velocity' which only depended on temperature and vanished at the lambda-point. The similarity with superconductivity was so inviting that it suggested a comparison experiment which could easily be performed with two concentrically arranged beakers (figure 11·6b). When the double beaker was lifted out of the liquid, the film had to run from the inner one into the outer beaker and from there into the bath. It turned out that there was *no* level difference between the two beakers, that means no difference in potential energy. This is quite analogous to the lack of electrical potential along a superconductive wire. Unfortunately it is much more difficult to make also a persistent current in liquid helium, but recent experiments indicate its existence.

With the discovery of film transfer we have come to the end of the hectic series of observations which was interrupted by the outbreak of war in 1939, except for the work in Moscow which continued for another two years. Let us therefore turn from the bewildering maze of experimental observations to the attempts which were made to interpret them. After his success with the superconductors, Fritz London turned to the problems of liquid helium and his first approach, made while he was still at Oxford, was based on a comparison with a crystal, a phase of thinking much in vogue at the time. He then accepted a professorship at the Sorbonne where the first news of the strange transport phenomena of He II reached him. In the same notable volume 141 of *Nature* he published his daring and exciting new theory which has remained controversial to this day. From his earlier interpretation, based on a crystal, he had changed to the other extreme of an ideal gas. Here again a long forgotten paper provided the clue.

Einstein, continuing his work on gas degeneracy which we discussed in chapter 7, arrived at a most astonishing conclusion. His calculation showed that an ideal gas obeying Bose statistics must undergo a curious change when cooled to a very low temperature. Then a point will be reached where some of the particles must 'condense'. However, the condensation predicted by Einstein does not result in a crystal since it will not take place in the space of positions but in that of velocities. To what kind of phenomena this hypothetical velocity condensation might give rise never became clear. Two years later doubts were raised about the validity of Einstein's conclusions and since, in any case, none of the known gases obeying Bose statistics appeared likely to become degenerate, the whole matter was forgotten.

London went back to Einstein's original paper and then calculated the specific heat due to such a condensation. It showed a peak which, however, did not resemble the lambda-point but, as London rightly remarked, liquid helium was far removed from an ideal gas. On the other hand, the low density due to zero point energy confers on the fluid some gas-like aspects. Altogether, London was extremely cautious, treating his idea as nothing more than a tentative suggestion and refraining from any attempt at explaining the transport effects. There is merely a short reference at the end, saying that his model might be of interest in that respect. Nevertheless, young theoreticians tend to rush in where experienced professors fear to tread. London had discussed his work with Lazlo Tisza, a Hungarian working at the Collège de France, and Tisza had drawn some remarkable conclusions which he published shortly afterwards, still in the same volume of *Nature*.

He applied, boldly and literally, London's suggestion to the observations on He II. As the liquid is cooled below the lambda-point, it will, according to Tisza, split up into two parts – the normal and the superfluid. The normal component is identical with He I but the superfluid one consists of 'condensed' atoms. It makes its first appearance at 2·2 K and on further cooling grows in quantity at the expense of the normal component until, at absolute zero the whole liquid is made up of superfluid. Thus, in this so-called 'two-fluid model' He II is regarded as a mixture of the normal and superfluid components in which the relative concentration varies with temperature (figure 10·7). It was in the physical properties of this mixture, as postulated from his model, where Tisza scored a remarkable triumph.

The superfluid component will be free of friction since it has no energy to be dissipated, and it will therefore lead to quite new hydrodynamical properties. The normal component, on the other hand, will behave exactly like He I. Taking the example of the viscosity measurements, it is obvious that the normal fluid will

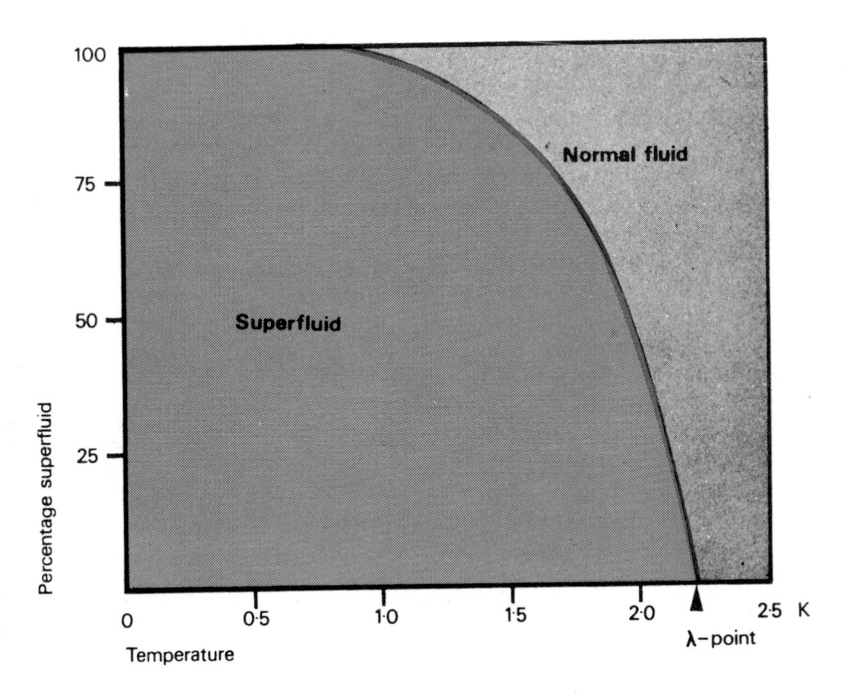

damp the motion of the oscillating disc and the drop observed in the viscosity below the lambda-point merely shows that the concentration of normal component diminishes. However, when using narrow slits or tubes, the normal component will hardly pass through them at all, but the superfluid will flow through without friction.

Equally, an explanation can be given for the thermo-mechanical effect (figure 11·5). As heat is supplied to the vessel on the right, superfluid is being transformed into normal and more superfluid passes through the capillary from left to right in order to compensate for the difference in concentration. The other way of equalising this difference, by normal fluid passing from right to left, is, however, impossible since it is stopped by friction in the capillary. The overall result is a net flow of liquid in the direction towards the heat. Reversing the process and forcing He II through a capillary will result in a cooling because only superfluid can pass. Accordingly, the concentration of normal fluid on the left side will be lower and this corresponds to a lower temperature.

The two-fluid model even offered an explanation for the high heat conductivity which Kapitza already had suspected to be a convection process. Tisza pointed out that since temperature changes always mean changes in concentration, the energy provided must be that which is required to excite helium atoms from zero energy to the He I state. Thus, at any convection the heat carried would not merely be the specific heat of the liquid but this high excitation energy.

When Tisza's paper was published, London was at first furious because he deplored the rash use of his own cautious suggestion. He also saw more clearly than anybody else the physical impossibility of admitting two fluids made up of the same species of atoms which by definition must be indistinguishable. Beyond this Tisza had postulated properties for the superfluid which did by no means follow from Bose–Einstein condensation. On the other hand, Tisza had been so incredibly successful in clearing up the tangled mess of experimental results by providing an interpretation which was at least logical that even London had to consider it as significant. Moreover, Tisza had used his model to make predictions and, as it turned out, these came true. The first was the mechano-caloric effect which he foresaw before the experiments were published. As mentioned above, this effect could be expected simply as a thermodynamic reversal of the thermo-mechanical one, and the prediction was a fairly safe one. The second prediction went much farther. He pointed out that a heat impulse fed into He II, being a momentary rise in concentration of normal fluid, should travel through the liquid as a thermal wave.

When Tisza's second paper in which he had made this prediction became available, war had broken out and, except in Russia, work on liquid helium had stopped. Owing to poor communications the Russian papers were slow in arriving. Even so, in the end we were able to make out what they had achieved before they, too, were overtaken by war. First of all, there was a long paper by Kapitza, written in 1940, which contained a large number of experiments, some of them very pretty. One of the most impressive ones had been designed to elucidate the heat conduction process. Heat was fed electrically into a double walled glass bulb which communicated with the helium bath through a tube (figure 11·8). Opposite to the orifice of the tube a vane was suspended so that it acted as an indicator for flow. When the heater was switched on, the vane was repelled by a stream issuing from the tube. This could be explained with the two-fluid model as normal fluid being produced at the heater and being ejected through the tube while a compensating, but frictionless, stream of superfluid flowed into the bulb. However, neither this explanation nor Tisza's papers are mentioned and they were evidently unknown. Instead, Kaptiza tried to visualise a counter-current running along the walls and in the middle of the tube.

In Kapitza's second paper, seven months later, he corrected this explanation and interpreted his observations in terms of the two-fluid model. Tisza, whose first paper had evidently arrived in the meantime, is mentioned, but Kapitza ascribed the true explanation to a theory by Lev Davidovich Landau which was published at the same time. Landau's theory, for which he received the Nobel Prize twenty years later, provided the much needed physical reason for the unreasonable success of Tisza's two-fluid model. According to Landau there exists just one fluid; liquid helium. As the temperature of the liquid is raised from absolute zero, thermal energy is provided in the form of vibrational quanta, the so-called phonons. These quantised vibrations of the helium atoms travel through the liquid in a manner which vaguely resembles the passage of light quanta, the photons, through space. In fact, the phonons are often described as 'quasi-particles' since they can be treated mathe-

11·8 Kapitza's demonstration of counterflow in He II.

matically in the same way as particles. The particle nature of the phonons is especially evident in the case of liquid helium. In Kapitza's experiment they are created at the heater and travel through the tube, finally impinging on the vane to which they impart their energy of motion. In short, Tisza's normal component corresponds to the quasi-particles travelling through the matrix of the fluid while the matrix itself is the superfluid.

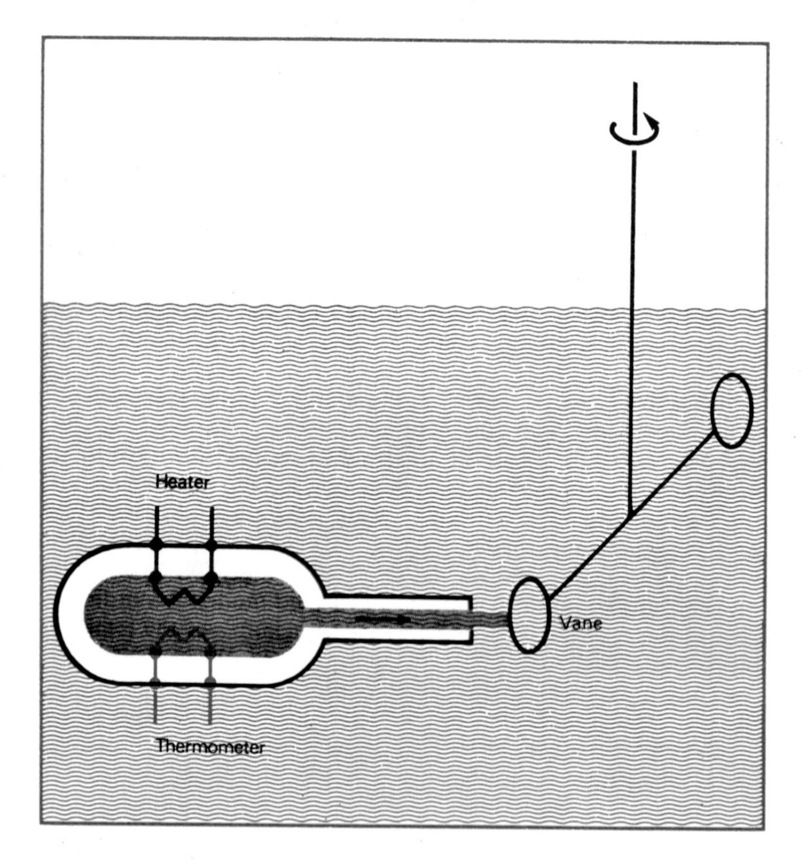

Landau's theory is not based on Bose–Einstein condensation although Bose statistics seem to be important. Essentially, the theory is meant to describe conditions close to absolute zero and it does not provide an explanation for the lambda-point. In addition to phonons, which are well known from the theory of solids, Landau postulated the existence of 'rotons', another type of thermal excitation which he introduced as elementary quanta of vortex motion. Various experiments made in the last decades have shown without doubt that such excitations exist in helium, but their actual nature is still something of a mystery.

Apart from furnishing a convincing and elegant interpretation of He II, the Landau theory has provided us with an interesting example of the way in which physics works. While evidently Tisza's first paper had reached Moscow by 1941, the second one had not. On the basis of his own theory Landau, too, had postulated a new form of wave motion in liquid helium but in his formalism it was similar to the acoustic phenomena and he therefore called it 'second sound'. In fact the first, unsuccessful, attempts at detecting it were made with acoustic equipment. It was only after E. M. Lifshitz re-interpreted Landau's formula, that the thermal nature of the waves was recognised. Their existence was demonstrated in 1944 by Peshkov and a later extension of the observations to temperatures below 1 K showed that Landau's, and not Tisza's, approach was the correct one.

In one respect the Landau theory gave the wrong prediction and that was the critical velocity of superflow. The fact that frictionless flow breaks down when a certain velocity of the superfluid is exceeded can be explained by the transformation of mechanical energy into heat, i.e. through the creation of phonons or rotons. Unfortunately, the experiments showed that superfluidity breaks down at velocities much smaller than those required to create these quasi-particles. Some other way in which the liquid can dissipate mechanical energy had to be found.

The fault really lay with the properties which the Landau theory ascribed to the superfluid matrix, and these excluded turbulence.

In fact, turbulence in the superfluid had been observed quite frequently, but somehow its significance was overlooked. It is now clear that the dependence of the heat conduction on the heat current originally observed in Cambridge in 1937, but then erroneously disbelieved, is evidence of turbulence. So was one of the Kapitza experiments in 1940. The key to an understanding of these phenomena was provided in 1949 by Onsager, but his important statement was merely made as a remark in a discussion at a conference in Florence and thus escaped notice.

Only in 1955 when Feynman in California devoted a long paper to Onsager's idea did its full significance for the explanation of superfluid phenomena begin to be realised. Onsager and Feynman postulated that the superfluid must be capable of forming eddies and that these large vortices, each containing an enormous number of atoms have to be quantised. Dissipation of energy can then occur when the vortices interact with phonons and rotons. Since so many atoms participate in the formation of these large scale quantum eddies, the energy per atom is much smaller than that needed for the creation of a single phonon or roton. In this way the vortices allow energy to be abstracted from the flow of the superfluid at velocities much smaller than those predicted by the Landau theory.

Circulation of the superfluid has been mentioned earlier in this chapter when, in conjunction with the double beaker experiment, we discussed the possibility of persistent fluid motion in He II. Quite a number of experiments carried out in the last decade make such large scale quantum circulation seem probable, although none are quite as conclusive as one would like them to be. Even so, nobody doubts seriously the existence of the phenomenon. However, frictionless flow and persistent currents are not the only features in which He II resembles a superconductor. As at the transition of a metal to the superconductive state, there occurs in helium also a rapid drop in the entropy. Moreover, superfluid flow, just like a persistent supercurrent, is distinguished by zero entropy, and we shall return to this point later.

Besides the usual type of helium atom whose nucleus consists of two protons and two neutrons, there exist others with only one neutron. This means that they have an odd number of spins and therefore must obey Fermi–Dirac and not Bose–Einstein statistics. This makes a comparison of the two isotopes particularly intriguing but, unfortunately, only every ten-thousandth helium atom is a light one. However, it is an ill wind that blows nobody any good, and the nuclear industry which makes atomic bombs also produces the light helium isotope. When this became clear, there arose the exciting prospect of investigating liquid light helium and finding out whether it would be superfluid. By 1949 enough light helium had become available to liquefy a tiny drop of it, and, just to spite the most fantastic prophecies made by many theoreticians, the experiment was a triumph for van der Waals. Using the old law of corresponding states to which a correction for the zero point energy had been added, de Boer of Amsterdam predicted correctly the vapour pressure curve of the new liquid with amazing accuracy. Its boiling point is one degree lower than that of ordinary helium.

Again, using the law of corresponding states, the lambda-point for the light isotope (we shall call it ^{3}He as distinct from the ordinary ^{4}He) had to be expected at about 1·5 K but no anomaly in the specific heat and no superfluidity was found at this or considerably lower temperatures. For the moment it seemed that Fritz London's explanation in terms of the Bose–Einstein condensation had been right. Subsequent work, however, has yielded many surprises in the behaviour of ^{3}He and it now appears that these may provide much profounder revelations on the nature of quantum fluids than we had suspected so far. However, before entering upon a discussion of these, we must for a moment return to the curious phase separation (figure 5·5) which we then quoted as the most striking example of the third law of thermodynamics. As mentioned then, a liquid mixture of the two isotopes will cease to be a homogeneous fluid when cooled below 0·8 K and split up into two separate phases, one being richer in ^{4}He which is superfluid and the other rich in ^{3}He which remains normal (figure 11·9).

11·9 Diagram showing the phase separation of liquid ^3He–^4He mixtures near absolute zero.

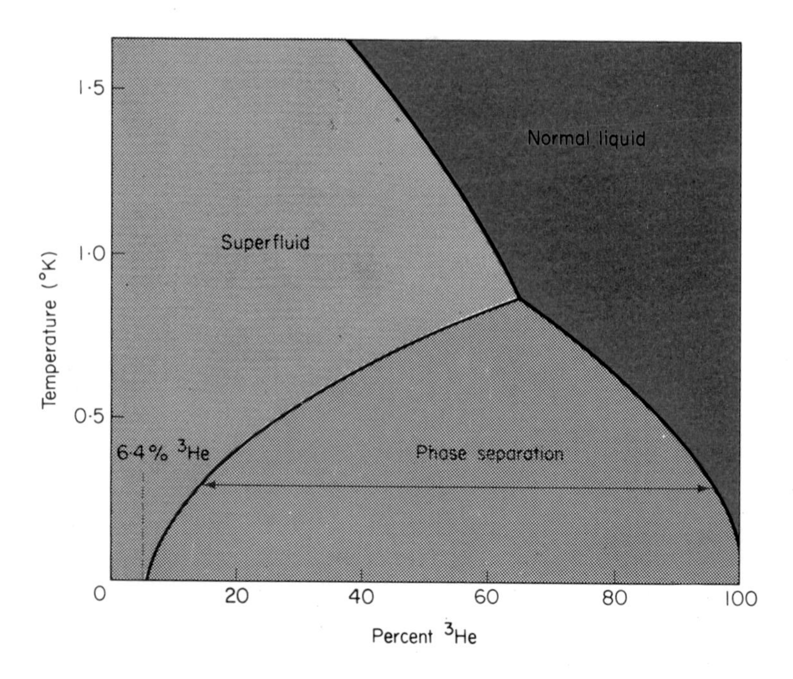

Even before the discovery of the phase separation, Heinz London suggested in 1951 that the entropy increase of mixing liquid ^3He and ^4He adiabatically should result in an appreciable cooling. The feasibility of this method was increased by the spontaneous phase separation and the first successful dilution refrigerator was operated in 1965 by Taconis and his colleagues in Leiden. Since then it has been extensively used and is now commercially available. The apparatus essentially consists of a closed circuit (figure 11·10) through which the working substance, a mixture of ^3He and ^4He is circulated by means of a pump (P). It is then condensed by contact with a bath of ^4He at 1·3 K (C) and further cooled in a heat exchanger (E) which brings it into a mixing chamber (M). In operation M becomes the coldest part of the cycle at about 0·1 K

11·10 Flow diagram of the helium dilution refrigerator. For explanation see text. ^{3}He rich phase pink, ^{4}He rich phase red.

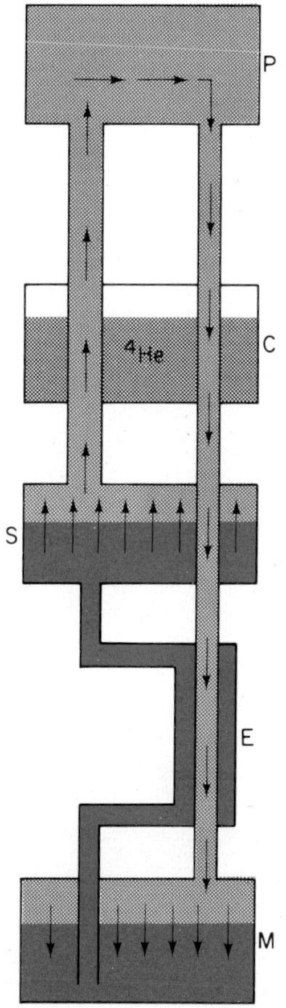

and it contains the two separated liquid phases with the lighter one, rich in ^{3}He on top. ^{3}He atoms from it now dissolve in the ^{4}He rich phase below and this process which resembles the evaporation of

11·11 Entropy diagram of liquid and solid ^{3}He. Solidification of the liquid phase by adiabatic compression leads to a lowering of temperature ($T_i \rightarrow T_f$).

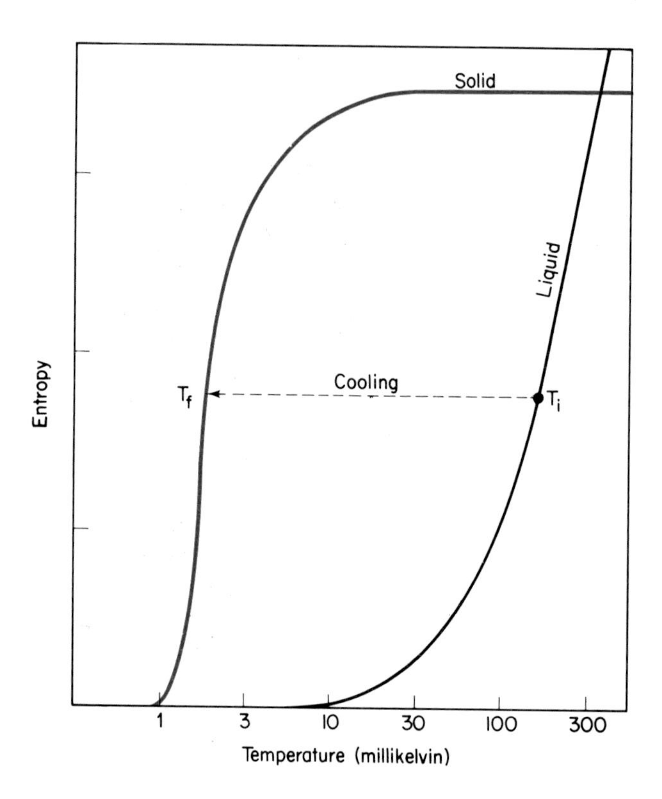

a gas produces the cooling. Experiments can now be carried out by attaching thermally the substance under investigation to M. The returning fluid now passes into a still (S) which is kept by a heater at 0·6 K from which ^{3}He is evaporated, passing back into the pump and leaving the ^{4}He rich fluid behind. As mentioned in chapter 8, the dilution refrigerator is increasingly used as the pre-cooling stage for nuclear refrigeration.

Returning now to the properties of pure ^{3}He, we have already seen that in its critical data it closely resembles the heavy isotope; its critical point is 3·32 K at a pressure of 1·15 atm. Like ^4He, the light isotope has high zero-point energy and under normal pressure remains in the liquid state down to absolute zero, solidifying only under a pressure of 34 atm. Here, however, the similarity ends and we now have to take account of the fact that, unlike the heavy isotope which obeys Bose statistics, ^3He is a fermion with a resultant unsaturated spin. It is the ordering of the spins which determines the entropy functions of liquid and solid ^3He which show (figure 11·11) that the spins in the solid phase remain disordered down to about 0·003 K whereas in the liquid the entropy decreases more rapidly. The behaviour of a Fermi liquid has been studied by Landau who showed that there is a close similarity with the normal electrons in a metal.

Even earlier, in 1950, Pomeranchuk in Moscow speculated on these entropy functions and proposed that adiabatic compression of liquid ^3He into the solid phase would result in cooling. It was, however, realised that the method, while feasible, would encounter appreciable practical difficulties and it was only fifteen years later that helium was first cooled with the Pomeranchuk method by Anufriev and in recent years temperatures of about 0·001 K have been reached by adiabatic solidification of ^3He. In fact, it was by using this method that the first indication of superfluidity in ^3He was obtained in 1972.

In the course of his work, Pomeranchuk predicted that the temperature (0·3 K) where the liquid entropy falls below that of the solid, the melting curve should pass through a minimum. Later this was, in fact, found to be correct (figure 11·12) and we see here that the magnetic properties of ^3He alter the thermodynamic functions when compared with ^4He. The most important difference between the isotopes appeared to be the failure of ^3He to become superfluid at about 1·5 K and at first it was thought that ^3He remains non-superfluid down to absolute zero. Subsequent theoretical work suggested that a new phase might appear at about 0·1 K. When eventually this temperature was reached, however,

11·12 Phase diagram of condensed ^{3}He. The melting curve has a minimum at about 0·3 K and in addition to the Fermi liquid which is stable above 0·001 K there exist two liquid modifications, *A* and *B*, the latter being superfluid.

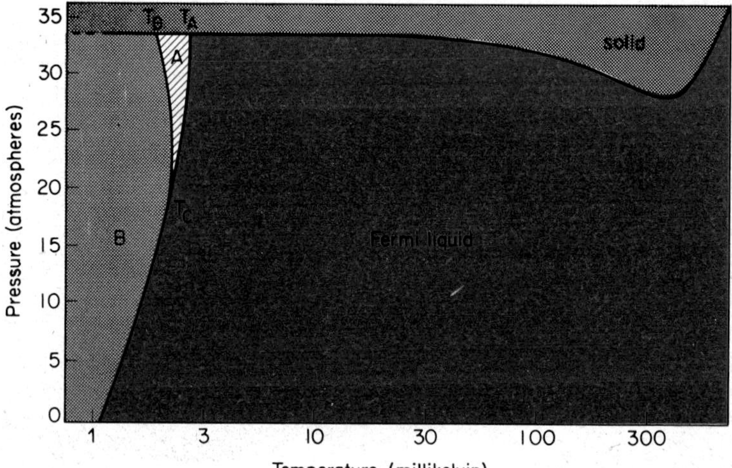

nothing was found, and from then onward a somewhat amusing contest developed between the theoreticians discovering further reasons why the transition should occur at still lower temperatures and the experimentalists finding nothing when they got that far. Then in 1964 Peshkov in Moscow announced the discovery of a specific heat anomaly at 0·005 K but this observation was hotly disputed by workers in Urbana. Eight years later such a transition was indeed found but at less than half this temperature and in retrospect one wonders whether Peshkov's discovery was right after all but that his temperature measurement was incorrect. Nobody will ever know nor care since the startling discoveries of the following years have now completely overshadowed this earlier work.

One of the groups pioneering Pomeranchuk cooling, working at Cornell University, perfected the method to an extent that they could follow the melting curve of ^{3}He down to less than 0·002 K. In 1972 Osheroff, Richardson and Lee noted that on cooling they passed two transition points at 0·0026 K and 0·002 K, indicating two new liquid phases. These observations triggered off a veritable avalanche of research in this very difficult field. The Cornell team, using Pomeranchuk cooling, had been forced to go along the melting curve and now the question as to the diagram of state at lower pressures had to be settled. For this purpose nuclear refrigeration had to be employed but the energetic pursuit of the problem has by now yielded a fairly detailed and reliable picture of the behaviour of liquid ^{3}He down to very low temperatures. The normal Fermi liquid which is the stable phase at all higher temperatures (figure 11·12) vanishes under zero pressure at a little above 0·001 K. Below this temperature a phase which has been named ^{3}He *B* is stable at low pressures. However, as we have just seen, at the melting curve there exist two transition points T_A and T_B leading to curves which merge at a tricritical point T_C at about 0·0024 K and 18 atm. These curves, together with the melting curve enclose another liquid phase ^{3}He *A*.

Thus the diagram of state of ^{3}He turns out to be more complex than that of the heavy isotope and again we have to look for the magnetic properties as the relevant feature. We cannot in this book go into any detail of the magnetic properties of the three ^{3}He phases but the important feature is that on passing from the normal Fermi fluid to the *A* liquid the viscosity diminishes greatly while at the transition to the *B* liquid true superfluidity is established. As in liquid ^{4}He the observed value of the viscosity in ^{3}He depends on the method of measurement but it appears that the *A* liquid is some sort of half-way house between the normal Fermi liquid and the fully superfluid *B* liquid. Again, over the whole region of temperature and pressure the thermodynamic properties vary with the strength of the magnetic field, presenting an exceedingly complex picture, a discussion of which is well beyond our scope.

Such attempts at a theoretical explanation which have been made to interpret these newly discovered phases of liquid ^{3}He closely follows the Bardeen, Cooper and Schrieffer theory of superconductivity. Again an energy gap and pairs of ^{3}He atoms in place of electrons, are invoked but with the difference that the ^{3}He pairs can each have opposite (↑ ↓) or parallel (↑ ↑) spin. A further variation is introduced by the possibility that parallel spin pairs can point in different directions and the *A* liquid has been visualised as a mixture of parallel spin pairs of opposite direction (↑ ↑, ↓ ↓). The *B* fluid, on the other hand, is being considered a true superfluid with pairs of opposite and equal spin and momentum (↑ ↓) in close similarity to the superconductive electron fluid.

Thirty years ago when I first drew attention to the basic similarity between the superfluidity of liquid helium and the superconductivity of electrons, physicists were far from convinced. It appeared rather too strange that helium atoms obeying Bose statistics and metal electrons which are fermions should show similar behaviour. Since then things have changed very much. The discovery of the Cooper pairs brought about a much closer similarity in the nature of the two types of carrier of superfluidity. Finally the recently discovered superfluidity of ^{3}He, again fermions, gives further strong support for this 'frictionless' state of aggregation of matter. This form of description chosen thirty years ago has now assumed a much wider and more profound significance since superfluidity appears to be a quite general feature of matter at low temperatures.

If anything, the different nature of the various superfluid carriers, but leading to the same aspects, strengthens rather than detracts from the universal nature of the phenomenon. There is an interesting parallel with the better known aggregation of solids. If we were not so very familiar with the existence of the solid state, we would be equally surprised that the various atoms and molecules, exerting different types of forces upon each other, always form crystals. Here again, the aspect of a crystal and its basic properties are much the same whether it is made up of copper atoms or hydrogen molecules or of such complex units as proteins.

The crystals with their rigidity and their sharp melting points owe this common aspect to the regular array of their constituent particles. The third law of thermodynamics demands that, as absolute zero is approached, the entropy must tend to zero, which means the substance must acquire an orderly pattern. The only orderly way in which a set of particles can be arranged side by side is in a regular array; they *must* form a crystal. I have purposely qualified this very general statement by the words 'side by side' in order to make it clear that we have only dealt with the regular pattern in the space of positions. This is the only type of space which we can comprehend on the basis of everyday experience, but it does not mean that there is no other space in which particles can form an orderly pattern.

As we have seen in chapter 5, 'before and after' is as important in the laws of physics as 'side by side'. Indeed the basic statement of quantum physics, the uncertainty principle, expresses Planck's constant through a combination of position and velocity. Neither of them has significance by itself; like lock and key they only acquire meaning when used together.

There is clearly no point in trying to decide whether in the description of the physical world position or velocity is of greater importance. They are both necessary and fully equivalent. Our discussion of gas degeneracy in chapter 7 has shown that the space of velocities is as significant as that of positions. It thus appears obvious that this equivalence in importance must also extend to the patterns of order in the two spaces. The fundamental features of the crystalline state are due to the fact that if is the most orderly pattern in position space. One therefore suspects that the unique and consistent aspect of the superfluids arises from an equally orderly pattern in the space velocities. The true nature of this order pattern remains obscure but it must appear significant that the strange properties which we observe in the superfluids are all concerned with states of motion.

When the first edition of this book was published ten years ago our idea that the superfluids might be a true counterpart to the crystal, being 'condensed' in velocities instead of position, was still a far-reaching assumption. However, with the discovery of superfluidity in liquid ^{3}He the emphasis has changed and instead of regarding the frictionless state of aggregate matter as an exception it now begins to look as if any assembly of freely mobile particles which at very low temperatures does *not* become superfluid may be an oddity. Moreover, there is now also the lingering suspicion that so far only part of the third law of thermodynamics may have been revealed to us. Nernst was, of course, right when postulating that at absolute zero the entropy S must be zero (at $T=0$; $S=0$) but he did not say how this ultimate state will be reached. For the normal electron gas as well as for the Fermi liquid, the entropy approaches absolute zero linearly and dS/dT remains finite. However, this state of affairs changes when the metal electrons or ^{3}He become superfluid. Then, as far as we know at present, not only S itself vanishes but dS/dT too.

Finally, there is another aspect to this apparent universality of the superfluid state of aggregation. The fact that in our world of observation it is tucked away into the realm of lowest temperatures does not mean that this will be the case for extra-terrestrial conditions. In certain stars, as for instance the white dwarfs, although the temperature is very high, the density is enormous. In such an assembly of freely moving particles of matter, degeneracy may force them into momentum condensation, resulting in phenomena of superfluidity. Perhaps this is the most fitting way to end a book on low temperatures. While on earth the superfluids are hidden at the lowest temperatures, the study of this obscure region may have revealed to us the most prevalent state of condensed matter in the universe. In fact, when compared with superfluidity, the solid state of crystalline aggregation may turn out to be a rare and quaint exception confined to highly unrepresentative conditions, such as the surface of this planet.

Bibliography

If a book has been published both in Britain and in the United States both publishers are listed, the British one being named first. Dates are of first publication.

J. F. Allen, ed., *Superfluid Helium,* Academic Press, London and New York, 1966.

K. R. Atkins, *Liquid Helium,* Cambridge University Press, 1959.

C. A. Bailey, ed., *Advanced Cryogenics,* Plenum Press, London and New York, 1971.

D. S. Betts, *Refrigeration and Thermometry below One Kelvin,* Sussex University Press, 1976.

N. N. Bogoliubov, ed., *The Theory of Superconductivity,* Gordon and Breach, New York, 1962.

E. F. Burton, H. G. Smith and J. O. Wilhelm, *Phenomena at the Temperature of Liquid Helium,* Chapman and Hall/Reinhold, 1940.

G. Claude, *Air Liquide, Oxygène, Azote,* Paris, 1909.

S. C. Collins, R. L. Cannaday, *Expansion Machines for Low Temperature Processes,* Oxford University Press, 1958.

A. J. Croft, *Cryogenic Laboratory Equipment,* Plenum Press, New York, 1970.

J. G. Daunt, ed., *Helium Three,* Ohio State University Press, 1960.

A. J. Dekker, *Solid State Physics,* Macmillan/Prentice Hall, 1957.

Sir James Dewar, *Collected Papers,* 2 vols, C.U.P./Macmillan, 1927.

F. Din and A. H. Cockett, *Low Temperature Techniques,* Newnes, London, 1960.

W. Frost, *Heat Transfer at Low Temperatures,* Plenum Press, New York, 1975.

C. G. B. Garrett, *Magnetic Cooling,* Chapman and Hall/Harvard U.P., 1954.

H. J. Goldsmid, *Thermoelectric Refrigeration,* Plenum Press, New York, 1964.

E. S. R. Gopal, *Specific Heat at Low Temperatures,* Plenum Press, New York, 1966.

C. J. Gorter, ed., *Progress in Low Temperature Physics,* 4 vols, North-Holland, Amsterdam, 1964.

Handbuch der Physik, vol. XV, Springer Hamburg.

H. Haken and M. Wagner, ed., *Cooperative Phenomena,* Springer, Berlin, 1973.

W. Heisenberg, *Die Physikalischen Prinzipien der Quantentheorie,* S. Hirzel, Stuttgart, 1930.

F. E. Hoare, L. C. Jackson and N. Kurti, ed., *Experimental Cryophysics.* Butterworth, London, 1961.

C. M. Hurd, *The Hall Effect in Metals and Alloys,* Plenum Press, New York, 1972.

L. C. Jackson, *Low Temperature Physics,* 5th edn, Methuen/Wiley, 1962.

W. H. Keesom, *Helium,* Elsevier, Amsterdam, 1942.

W. E. Keller, *Helium-3 and Helium-4,* Plenum Press, New York, 1969.

C. Kittel, *Introduction to Solid State Physics,* Wiley, New York, 1956.

C. Kittel, *Solid State Physics,* Wiley, New York, 1962.

J. A. van Lammeren, *Technik der tiefen Temperaturen,* Springer, Hamburg, 1941.

C. T. Lane, *Superfluid Physics,* McGraw-Hill, New York, 1962.

M. von Laue, *Theory of Superconductivity,* trans. Lothar Meyer and William Band, Academic Press, London and New York, 1952.

E. M. Lifshits and E. L. Andronikashvili, *Supplement to 'Helium',* Consultants Bureau, London and New York, 1959.

F. London, *Superfluids,* 2 vols, Chapman and Hall/Wiley, 1954.

H. A. Lorentz, ed., *Het Natuurkundig Laboratorium der Rijksuniversiteit te Leiden in den Jaren 1904–1922,* E. Ijdo, 1922.

O. V. Lounasmaa, *Experimental Principles and Methods below 1 K,* Academic Press, London and New York, 1974.

E. A. Lynton, *Superconductivity,* Methuen/Wiley, 1962.

G. T. Meaden, *Electrical Resistance of Metals,* Plenum Press, New York, 1965.

K. Mendelssohn, *Cryophysics,* Interscience, London and New York, 1960.

K. Mendelssohn, ed., *ICEC Proceedings 1967, 1968, 1970, 1972, 1974,* IPC Science and Technology Press, London.

K. Mendelssohn, ed., *Progress in Cryogenics,* 4 vols, Heywood/Academic Press, 1964.

K. Mendelssohn, *The World of Walter Nernst,* Macmillan, London, 1973.

W. Nernst, *Die Theoretischen und Experimentellen Grundlagen des Neuen Waermesatzes,* W. Knapp, Düsseldorf.

W. Nernst, *Theoretische Chemie,* F. Enke, Stuttgart.

V. L. Newhouse, *Applied Superconductivity,* Wiley, New York, 1964.

Physikalische Gesellschaft Zurich, ed., *Vorträge über Supraleitung,* Birkhäuser, Basel, 1968.

J. L. Olsen, *Electron Transport in Metals,* Interscience, London and New York, 1962.

A. B. Pippard, *Classical Thermodynamics,* C.U.P., 1964.

M. Planck, *Thermodynamik,* W. de Gruyter, Berlin.

M. Planck, *Waermestrahlung,* J. A. Barth, München.

A. C. Rose-Innes, *Low Temperature Techniques,* The English University Press, London, 1964.

A. C. Rose-Innes and E. H. Rhoderick, *Introduction to Superconductivity,* Pergamon Press, Oxford, 1969.

H. M. Rosenberg, *Low Temperature Solid State Physics,* O.U.P., 1963.
M. and B. Ruhemann, *Low Temperature Physics,* C.U.P./Macmillan, 1937.
E. M. Savitskii, V. V. Baron, Y. V. Efimov, M. I. Bychkova and L. F. Myzenkova, *Superconductive Materials,* Plenum Press, New York, 1973.
R. B. Scott, *Cryogenic Engineering,* Van Nostrand, New York, 1959.
R. B. Scott, ed., *Technology and Uses of Liquid Hydrogen,* Wiley, New York, 1962.
F. Seitz, *The Modern Theory of Solids,* McGraw-Hill, New York, 1940.
D. Shoenberg, *Superconductivity,* C.U.P., 1952.
F. E. Simon, N. Kurti, J. F. Allen and K. Mendelssohn, *Low Temperature Physics; Four Lectures,* Pergamon, London, 1952.
C. F. Squire, *Low Temperature Physics,* McGraw-Hill, New York, 1953.
H. N. V. Temperley, *Changes of State,* Cleaver Hume/Interscience, 1956.
R. W. Vance and W. M. Duke, ed., *Applied Cryogenic Engineering,* Wiley, New York, 1962.
G. K. White, *Experimental Techniques in Low Temperature Physics,* Clarendon Press, Oxford, 1968.
D. A. Wigley, *Mechanical Properties of Materials at Low Temperatures,* Plenum Press, New York, 1971.
J. Wilks, *Liquid and Solid Helium,* Clarendon Press, 1967.
B. Yates, *Thermal Expansion,* Plenum Press, New York, 1972.
M. G. Zabetakis, *Safety with Cryogenic Fluids,* Plenum Press, New York, 1967.
M. W. Zemansky, *Heat and Thermodynamics,* 4th edn, McGraw-Hill, New York, 1957.
J. Ziman, *Principles of the Theory of Solids,* C.U.P., 1964.

Acknowledgments

The author is indebted to Mr. W. J. Green for information on Dewar and to those colleagues and scientific institutions who provided opportunities for parts of this book to be written at Lake Como (in summer) and in equatorial Africa (in winter).

Acknowledgment is due to the following for the illustrations (the number refers to the page on which the illustration appears). 196, figure 11.1 Clarendon Laboratory, Oxford University (photos Cyril Band); 60 (Cailletet) Académie des Sciences, Institut de France; (Wroblewski, Olszewski) Professor H. Niewodniczański; (Andrews) The Royal Society (photo John Freeman); (Pictet) Professor J. Muller (photo Jean Arland); (Dewar) The Royal Institution of Great Britain; (Kammerlingh Onnes) Kammerlingh Onnes Laboratory, Leiden University; (Giauque) Professor Giauque; (Einstein) Radio Times Hulton Picture Library; (Debye) Professor Debye; 68/9 The Royal Institution of Great Britain (photo John Freeman); 107 Professor V. Peshkov; 194 A. D. Little Inc., Cambridge, Mass.; 212 Professors Essmann and Rose-Innes; 230, 231 Dr. Catterall, U.K. Atomic Energy Research Establishment, Harwell, and Rutherford Laboratory; 233 Clarendon Laboratory and Oxford Instruments; 235 Drs. Garwin and Matisoo; 237 International Research and Development Co.; 239 Japanese National Railways; 240 Financial Times; 268 Professor Lounasmaa.

Index